F-105 THUNDERCHIEF

Walter J. Boyne Military Aircraft Series

F-22 Raptor
America's Next Lethal War Machine
STEVE PACE

B-1 Lancer
The Most Complicated Warplane Ever Developed
DENNIS R. JENKINS

B-24 Liberator
Rugged but Right
FREDERICK A. JOHNSEN

B-2 Spirit
The Most Capable War Machine on the Planet
STEVE PACE

F/A-18 Hornet
A Navy Success Story
DENNIS R. JENKINS

B-17 Flying Fortress
The Symbol of Second World War Air Power
FREDERICK A. JOHNSEN

F-105 Thunderchief
Workhorse of the Vietnam War
DENNIS R. JENKINS

B-47 Stratojet
Boeing's Brilliant Bomber
JAN TEGLER

F-105 THUNDERCHIEF

Workhorse of the Vietnam War

Dennis R. Jenkins

McGraw-Hill
New York San Francisco Washington, D.C. Auckland Bogotá
Caracas Lisbon London Madrid Mexico City Milan
Montreal New Delhi San Juan Singapore
Sydney Tokyo Toronto

Library of Congress Cataloging-in-Publication Data

Jenkins, Dennis R.
    F-105 Thunderchief : workhorse of the Vietnam War /
  Dennis R. Jenkins.
        p.   cm.
    Includes index.
    ISBN 0-07-135511-1
    1. Thunderchief (Fighter planes)   2. Vietnamese Conflict,
1961-1975–Aerial operations, American.   I. Title.
UG1242.F5 J4623   2000
623.7'464'0973—dc21
                                          00-027667

McGraw-Hill
A Division of The McGraw·Hill Companies

Copyright © 2000 by Dennis R. Jenkins. All rights reserved. Printed in the United States of America. Except as permitted under the United States Copyright Act of 1976, no part of this publication may be reproduced or distributed in any form or by any means, or stored in a data base or retrieval system, without the prior written permission of the publisher.

1 2 3 4 5 6 7 8 9 0   QK/QK   9 0 9 8 7 6 5 4 3 2 1 0 9

ISBN 0-07-135511-1

The sponsoring editor for this book was Shelley Ingram Carr, the editing supervisor was Sally Glover, and the production supervisor was Pamela Pelton. It was set in Utopia by North Market Street Graphics.

Printed and bound by Quebecor/Kingsport.

 This book is printed on recycled, acid-free paper containing a minimum of 50% recycled, de-inked fiber.

McGraw-Hill books are available at special quantity discounts to use as premiums and sales promotions, or for use in corporate training programs. For more information, please write to the Director of Special Sales, McGraw-Hill, Two Penn Plaza, New York, NY 10121-2298. Or contact your local bookstore.

Information contained in this work has been obtained by The McGraw-Hill Companies, Inc. ("McGraw-Hill") from sources believed to be reliable. However, neither McGraw-Hill nor its authors guarantees the accuracy or completeness of any information published herein, and neither McGraw-Hill nor its authors shall be responsible for any errors, omissions, or damages arising out of use of this information. This work is published with the understanding that McGraw-Hill and its authors are supplying information but are not attempting to render professional services. If such services are required, the assistance of an appropriate professional should be sought.

CONTENTS

THE WALTER J. BOYNE MILITARY AIRCRAFT SERIES — vii
FOREWORD BY WALTER J. BOYNE — ix
FOREWORD BY COLONEL CLARENCE E. ANDERSON, JR. — xi
PREFACE — xiii

1. A New Role—Tactical Nuclear Strike — 1
F-105A — 3
YF-105A — 5
North American YF-107A — 11

2. In Production at Last—The F-105B — 17
RF-105B/JF-105B — 35
F-105C — 38
Thunderbirds — 39

3. Real Thunderchiefs—The F-105D/F — 45
Thunderstick II — 58
RF-105D — 59
F-105E — 61
Finally, a Two-Seater: The F-105F — 64
F-105G — 68
F-105H — 68
Other Stillborn Thuds — 68

4. F-105D/F Operational Service—Vietnam and Thud Ridge — 73
SEA Modifications — 86
Electronic Countermeasures — 90
Commando Nail — 96
Combat Martin — 103

Combat Skyspot	104
The End	106

5. Wild Weasels—The F-105F/G — **111**

Wild Weasel II	113
Wild Weasel IA	115
Wild Weasel III	116
Two Very Special Missions	143
The End	148

6. The Details—Construction and Systems — **155**

Cockpit	155
Fuselage	157
Wings	159
Tail Surfaces	161
Landing Gear	162
Electrical System	163
Hydraulic System	163
Miscellaneous Systems	164
Avionics	165
Paint and Markings	167
Armament	169
Powerplant	170

Appendix A: Fire Control System Development	**179**
Appendix B: Acronyms	**185**
Appendix C: Serial Numbers	**189**
Appendix D: F-105 Losses in Southeast Asia	**195**
Appendix E: Thunderchief MiG Killers	**221**
INDEX	223

THE WALTER J. BOYNE MILITARY AIRCRAFT SERIES

The McGraw-Hill Companies is pleased to present the **Walter J. Boyne Military Aircraft Series.** The series will feature comprehensive coverage, in words and photos, of the most important military aircraft of our time.

Profiles of aircraft critical to defense superiority in World War II, Korea, Vietnam, the Cold War, the Gulf Wars, and future theaters, detail the technology, engineering, design, missions, and people that give these aircraft their edge. Their origins, the competitions between manufacturers, the glitches and failures and type modifications are presented along with performance data, specifications, and inside stories.

To ensure that quality standards set for this series are met volume after volume, McGraw-Hill is immensely pleased to have Walter J. Boyne on board. In addition to his overall supervision of the series, Walter is contributing a Foreword to each volume that provides the scope and dimension of the featured aircraft.

Walter was selected as editor because of his international preeminence in the field of military aviation and particularly in aviation history. His consuming, lifelong interest in aerospace subjects is combined with an amazing memory for facts and a passion for research. His knowledge of the subject is enhanced by his personal acquaintance with many of the great pilots, designers, and business managers of the industry.

As a Command Pilot in the United States Air Force, Colonel Boyne flew more than 5000 hours in a score of different military and civil aircraft. After his retirement from the Air Force in 1974, he joined the Smithsonian Institution's National Air & Space Museum, where he became Acting Director in 1981 and Director in 1986. Among his accomplishments at the Museum were the conversion of Silver Hill from total disarray to the popular and well-maintained Paul Garber Facility, and the founding of the very successful *Air & Space/Smithsonian* magazine. He was also responsible for the creation of NASM's large, glass-enclosed restaurant facility. After obtaining permission to install IMAX cameras on the Space Shuttle, he supervised the production of two IMAX films. In 1985, he began the formal process that will lead ultimately to the creation of a NASM restoration facility at Dulles Airport in Virginia.

Boyne's professional writing career began in 1962; since that time he has written more than 500 articles and 28 books, primarily on aviation subjects. He is one of the few authors to have had both fiction and nonfiction books on the *New York Times* best seller lists. His books include four novels, two books on the Gulf War, one book on art, and one on automobiles. His books have been published in Canada, Czechoslovakia, England, Germany, Italy, Japan, and Poland. Several have been made into documentary videos, with Boyne acting as host and narrator.

ABOUT THE AUTHOR

Dennis R. Jenkins has been a senior engineer/manager in the aerospace industry for more than 20 years. For corporations such as Lockheed Martin, he has worked on projects that included the Space Shuttle, the Ballistic Missile Early Warning System, X-33/VentureStar™, and the National Airspace System (FAA). The author of several books on air- and spacecraft, including *B-1 Lancer* (another volume in the *Walter J. Boyne Military Aircraft Series*), he has degrees in both computer engineering and R&D systems management.

FOREWORD BY WALTER J. BOYNE

The remarkable service life of the remarkable Republic F-105 Thunderchief can be considered a monument to overcoming contradictions. Originally designed to deliver nuclear weapons to the heartland of the Soviet Union, the F-105 was destined to fight a conventional war using, for the most part, conventional iron bombs. Equipped with a powerful engine and miniscule wings, its original mission called for it to fly straight and level at supersonic speeds. When it went to war it had to fly a hard-jinking flight path to avoid surface-to-air missiles and anti-aircraft fire, and, all too often, to engage in combat with the agile MiG fighters. The powerful Pratt & Whitney J75 engine delivered tremendous power but gulped fuel in enormous quantities, making multiple in-flight refuelings commonplace on long missions. Loaded with ultrasophisticated equipment intended to be maintained at well-equipped bases in temperate climates, it fought its war from newly created bases in Thailand, suffering continuous assault from high heat and humidity.

Despite all these contradictions, the thud, as it was affectionately called, proved to be a stellar performer, a champion in the most difficult arena of all, the metal-filled skies of North Vietnam. The F-105 is beloved by the pilots who flew it and by the men who maintained it, for its incredible rugged strength. Yet, as you'll see in the following pages, the conditions under which the thud fought were so severe that sheer attrition eventually forced it out of service.

Photographs and paintings of the F-105 can give an idea of its sculptural beauty, for it has the smooth lines and seductive curves that distinguished all Republic aircraft, right back to the P-35As, P-43s and P-47s. Yet to truly appreciate its beauty, one had to see it firsthand, to walk around it on the flight line and get the true impression of its enormous size. Then, as I had the opportunity to do so many times, I got another impression of its beauty when I watched it take off. It seemed almost too squat at the end of the runway, its wings encumbered with external sources. The afterburner would kick in with a roar, the thud begin to roll, slowly at first, and then accelerating at a tremendous pace, eating up great lengths of runway as it did so. Finally, it would be airborne, gear and flaps coming up, transforming itself into one of the most beautiful aircraft of the time.

When the F-105 prototype made its first flight on October 22, 1955, it represented the absolute epitome of modern aircraft design in every respect. Both wings and tail surfaces had an aggressive 45-degree sweep-back, the short wings were very thin, and, overall, it possessed the usual purity of line demanded by the great designer Alexander Kartveli. Ironically, this sought-for and achieved purity of line was not compatible with the new transonic and supersonic speed regime in which it was to be flown. Even as the original F-105 design was being laid out, Dr. Richard Whitcomb, working at the National Advisory

Committee for Aeronautics (NACA, later to become NASA) had developed his brilliant, Collier Trophy–winning concept of the area rule. In brief, Whitcomb had found that the drag reduction necessary for transonic and sonic speeds could only be achieved if careful consideration were given to the distribution of the combined cross sections of the wing and fuselage.

The prototype F-105s, like the prototype Convair F-102, were not capable of supersonic flight and had to be redesigned. This was anathema to Kartveli, who could not believe that bulges, no matter how carefully formulated, should be applied to his brainchild.

Yet the bulges were applied, and they worked, converting the F-105 into a wasp-waisted supersonic aircraft. Ironically, instead of impairing the looks of the F-105, the design changes actually improved its appearance, giving it a sensuous look that stood it well throughout its servce.

Although it was eight years after Yeager had broken the sound barrier, much still needed to be learned about supersonic flight, and tests resulted in many more modifications to the aircraft. One distinguishing if disconcerting aspect of the test program was that three of the first four Thunderchiefs had to make belly landings, a situation that was not repeated with production aircraft.

The thud was the workhorse of the Vietnamese War, reaching out on a daily basis for bases in Thailand to fly to Route Package 6. There the bombs were placed on carefully selected targets, under blindly unfair rules of engagement that gave every advantage to the enemy. The F-105 pilots accepted the situation and carried on with their mission, determined to fight the war as aggressively and effectively as they could.

The North Vietnamese were very capable enemies, well equipped with massive amounts of supplies from the Soviet Union and from China. They knew that the F-105 targets were concentrated in a very small area just as they knew that the self-imposed American rules of engagement allowed the choice of only a few possible routes to the target. They quite naturally concentrated their SAMs, antiaircraft, and MiG fields in areas that covered those routes.

Despite the many hazards, the thud pilots executed their missions on a daily basis, accepting their inevitable losses. As time passed, however, the enemy SAM defenses became ever more formidable, and it was necessary to find a way to neutralize them. The result was the creation of the Wild Weasel mission, using special crews and equpment to suppress the enemy air defenses.

The first Wild Weasel aircraft were North American F-100F Supersabres which had been equipped with hastily developed electronic gear to detect the SAM sites. Heroically flown, the F-100s proved the theories and developed the techniques, but casualties were high, with almost 50 percent of the force being lost in combat. More important, the F-100 lacked the performance to work with the much faster 105s. The answer, naturally enough, was the two-seat F-105F, equipped with the latest in defense suppression gear.

Unfortunately, there were relatively few qualified crews available, for a training program had not been set up at Nellis AFB until February, 1966. Fortunately, it was a superb program manned by veterans just returned from Southeast Asia, and it soon began furnishing well-qualified crews. The F-105 Wild Weasels proved to be an unqualified success, and their motto "First In, Last Out" symbolized not only their mission but their didication to their comrades.

By 1972, attrition and hard usage had caused the F-105s to be replaced by McDonnell Douglas F4 Phantoms, *except* in the Wild Weasel role, where the Thuds soldiered on, fighting an ever more sophisticated enemy. These were themselves eventually replaced by F4C Wild Weasel conversions.

The F-105 had started out its life as a nuclear bomber, one bemoaned for its expense and its initial difficulties. When combat came, however, the Republic Thunderchief was flown to glory by its pilots, who relished its speed, and were grateful for its rugged construction. Today one only has to talk to a former F-105 pilot to learn just what a superb airplane it proved to be.

FOREWORD BY COLONEL CLARENCE E. "BUD" ANDERSON, JR., USAF (RET.)

The Republic F-105 Thunderchief was just starting Air Force flight tests at Edwards Air Force Base when I was assigned there as the Chief, Flight Test Operations Division. It already had a poor reputation as a fighter, but nevertheless it was soon to be delivered to the Tactical Air Command squadrons. I flew an evaluation flight in the F-105 and had to agree that maybe we should have called it a bomber instead of a fighter. Indeed, it was designed to be just that, to deliver a nuclear bomb at low altitude. It was big, the largest single engine fighter ever built, and it flew a great deal like the earlier subsonic Republic jet fighters. They all flew well but were heavy, used a lot of runway for takeoff, and were not considered particularly agile. The straight-wing F-84s had been dubbed the Hog, and the swept-wing F-84Fs became the Super Hog. Now names like the Ultra Hog or even the Lead Sled were tagged onto the F-105. The name did not stick long.

The F-105 ran into all kinds of problems as soon as it was deployed. The F-105 flew well but had a rash of problems with about every aircraft system you could name. The Air Force "Thunderbirds" demonstration team converted to the F-105 and soon one had a structural failure and broke up during an aerial demonstration. Meanwhile, there were many other unrelated crashes of the F-105, resulting in the aircraft being grounded several times. Numerous corrective modifications were issued from Republic to fix the deficiencies. Because of the crashes, the F-105 became know as the "Thud," and not with affection either.

In mid-1965, I was assigned to the 18th TFW on the island of Okinawa, and much to my dismay I learned that the Wing was equipped with the F-105. During my tour with the 18th TFW, I would learn a lot about the Thud. The F-105s were in the middle of one of the latest safety modifications. We had aircraft with cracked wing spars standing on nuclear alert. The pilots had assigned targets in case of nuclear war and we would have flown them if necessary, but they were not flown in a day-to-day training environment until the spars were reinforced. Soon the F-105 fleet became quite reliable. Our squadrons started rotational deployments to Southeast Asia for combat missions over North Vietnam. I think that this tour was one of the busiest times in my Air Force career. The experience level was very high and, with the extensively modified F-105 finally becoming reliable, the Wing became very professional. We had a nuclear alert commitment, a full training schedule, Operational Readiness Inspections, squadrons on temporary combat duty in Thailand, and a battle damage repair and aircraft replacement mission. The more I flew the F-105 now, the more I was impressed. One thing I remember is that you didn't want to engage anyone in a close-in dogfight unless it was another F-105. However, it was an excellent bombing and gunnery platform. Outstanding accuracy could be achieved on the target ranges, and I was impressed that one could deliver a simulated nuclear weapon on instruments at a well-

During World War II, Bud Anderson flew a North American P-51 Mustang named "Old Crow." The name was also used on the F-105D that Anderson flew in Southeast Asia. Note the old crow insignia behind the boarding ladder. *(C. E. Anderson)*

defined radar target. The stories started coming back from the Vietnam War about the damage the F-105 could inflict. Further, it could take lots of damage and still get home or at least to a safe zone for ejection and rescue. The name Thud took on a new meaning for the F-105 pilots who now loved the Thunderchief for the job it was doing in the Vietnam War.

Later I was assigned to the 355th TFW in Thailand near the end of the Vietnam War and I saw firsthand how the F-105 performed in combat. I have great respect for the F-105 "Thud" and its air crews, especially those pilots who flew in the early part of the war on missions over North Vietnam.

Colonel Clarence E. "Bud" Anderson, Jr., USAF (Ret.) flew two combat tours during World War II, scoring 16.25 aerial victories during 116 combat missions. Anderson went on to Wright-Patterson AFB as a fighter test pilot, including participation in projects TIP-TOW and FICON, and later became Chief of Fighter Operations. While at the Air Force Flight Test Center at Edwards AFB, Colonel Anderson was assigned as the Chief of Flight Test Operations and later Deputy Director of Flight Test. Other duty included two tours at the Pentagon, commanding a squadron of F-86s in post-war Korea, commanding the 18th TFW with F-105s on Okinawa, and flying 25 combat missions in F-105s while commander of the 355th TFW in Thailand. After retiring from the Air Force in 1972, Anderson joined the McDonnell Aircraft Company and served for 12 years as Manager of the company flight test facility at Edwards AFB.

Anderson's combat aircraft have always been called "Old Crow," and he can still occasionally be seen flying a P-51 with that name. He has flown over 130 different types of aircraft and has logged over 7,500 flying hours. Colonel Anderson was decorated 25 times, including 2 Legion of Merits, 5 Distinguished Flying Crosses, the Bronze Star, 16 Air Medals, and the French Croix de Guerre. He is married to the former Eleanor Cosby, and they have two children and four grandchildren. In 1990 Colonel Anderson wrote an autobiography which has been described by the Historian of the Air Force as "the finest pilot memories of WW II." In that book, titled TO FLY AND FIGHT, General Chuck Yeager describes Anderson as "a mongoose, . . . the best fighter pilot I've ever seen."

PREFACE

The Republic F-105 Thunderchief ("... don't call it "Thud" unless you flew it...") has long been a favorite of mine. Even before I knew of its mixed history—all you hear about anymore is its Vietnam experience—the aircraft intrigued me. When you stand beside one it is hard to believe it only carries one person and is powered by a single engine. Although several modern fighters are larger and heavier, none look as massive as the F-105.

The F-105 gained its fame in the skies over Southeast Asia—carrying weapons it was not designed to use in a war it was not supposed to fight. The Thunderchief had been conceived to carry the new tactical nuclear weapons that the national laboratories were designing during the early 1950s. Its target was the Evil Empire of the Soviet Union during the height of the Cold War. It was meant to operate from well-equipped and maintained bases in Europe and Japan. Instead it found itself in the jungles of Thailand, surrounded by humidity, trying to fight a war by rules made up by politicians 7,000 miles away. Too many missions were flown with only one or two bombs per aircraft, simply so the politicians could count sorties. Too many crews never came back.

It was a cruel war. The F-105 holds the distinction of being the only American combat aircraft withdrawn from service simply because there were not enough of them remaining to be tactically useful. Over 20,000 combat missions were flown by Thunderchiefs in Southeast Asia, resulting in the loss of 331 aircraft, most of them to North Vietnamese antiaircraft fire. This was almost half of all combat-capable F-105s built. One hundred twenty-four Thunderchiefs were lost in 1966 alone, 91 of them to AAA. One hundred fifty-six crewmembers were listed as KIA or MIA. Two pilots won the Medal of Honor for their missions.

What is impossible to know is how many lives the F-105 saved. During Linebacker the Wild Weasels were responsible for protecting the B-52s, along with other aircraft, from surface-to-air missiles. The entire concept of the early Weasel is frightening. Present yourself as a target for a large flying "telephone pole," then turn and attack the launch site, hopefully without getting yourself killed in the process. But every missile fired at a Weasel, and every SAM site killed, was one less that could be fired at the bombers.

In 1999 there is not a single flyable Thunderchief, although at least 45 of them are in museums, available to be admired at one's leisure. One should take the leisure.

This book would not have been possible without the assistance of C. E. "Bud" Anderson, Tony Landis, Mike Machat, Terry Panopalis, Ken Neubeck, Larry Davis, Frederick A. Johnsen, Ed Rasimus, Douglas R. Thar (SAF/PAN), Jerry Langston, Archie DiFante at the Air Force Historical Research Agency, Dr. Don Klinko at Ogden ALC, Dr. Craig Luther at Sacramento ALC, Colonel Edward T. Rock, USAF (ret.), Bill Evans, Wesley B. Henry and Dave Menard at the Air Force Museum, Thomas Kreutzberg at Fairchild-Republic, TSgt. Gary Cox

Republic had a long, proud tradition of building front-line fighters for the Air Force. In May 1961 the company celebrated the 20th anniversary of the first flight of the P-47 Thunderbolt by posing a late-model P-47 with a brand-new F-105D Thunderchief. The F-105 would prove to be the last fighter produced by Republic. *(Republic Aviation via the Frederick A. Johnsen Collection)*

(USANG—184 TFG), MSgt. Tom Ferrell (USANG—113 CAM/MAQ), Eric Renth, Larry S. Johnson, and William J. Simone.

Special thanks go to Mick Roth, who wrote the Wild Weasel chapter contained within. Mick has spent years studying the history of the Wild Weasel, and I am deeply indebted for the amount of work he put into preparing this chapter. I am also grateful for the support of Walter J. Boyne, and the entire series would not have been possible without the efforts of Shelley Carr at McGraw Hill. On the personal side, thanks to Cyndy Thomas and my mother, Mary E. Jenkins, for putting up with the number of hours I spent on researching and writing.

—Dennis R. Jenkins
Cape Canaveral, Florida
November 1999

A New Role—Tactical Nuclear Strike

THUD., n, nickname affectionately applied to a species of Air Force fighter-bomber found in considerable numbers over North Vietnam; origin of the name, THUD; U.S. Air Force F-4 pilots' description of the sound made by the Republic F-105 Thunderchief when it hits the ground. Also known as "Lead Sled," "Ultra Hog," "Squash Bomber," and "Drop Forged by Republic Aviation."[1]

During the early 1950s, the Republic Aviation Corporation was busy with a $580 million backlog of production orders for the straight-wing F-84E/G Thunderjet and was beginning test flights of the swept-wing F-84F Thunderstreak. Republic had a long and proud history of producing Army and Air Force fighters and was determined to continue doing so.

At the time, Republic was actively working on only one new design. Since 1948, Republic had been developing the AP-44A[2] all-weather, high-speed, high-altitude interceptor. The Air Force had approved development of the XF-103 as part of WS-204A in September 1951, and the full-scale mock-up was inspected on 2 March 1953. The inspection resulted in the replacement of the aircraft's canopy with a flush cockpit that used a periscope and an 18-month extension to conduct additional studies into titanium fabrication, high-temperature hydraulics, and escape systems. The XF-103 used a novel Wright MX-1787 turbo-ramjet powerplant and was expected to achieve Mach 3 at 80,000 feet—if only briefly—on its way to an intercept. During July 1954 the Air Force ordered three prototypes, but progress was slowed by difficulties in fabricating titanium parts and major engine development problems. By early 1957 the program had been cut back to a single flight vehicle and two engines, and any possibility of large-scale production had passed. In September 1957 the Air Force finally canceled the program after deeming the aircraft simply too advanced for existing technology. More than $109 million had been spent on its development over 9 years. Clearly the F-103 would not provide Republic with much future work.[3]

So Republic needed a new product to maintain its business base with the Air Force. The later F-84 series were proving to be underpowered; as usual, powerplant performance had not kept up with the increased weight of added systems. The first of the Century Series fighters were beginning flight testing, and it appeared that their quantum leap in performance would make most previous fighters obsolete almost overnight. Perhaps more significantly, a small tactical nuclear weapon was on the drawing boards, and the Air Force had decided that the weapon should be carried internally by a fast fighter-bomber.

The Air Force did not have funding to study an improved F-84, so Republic took it upon itself to do so. Alexander Kartveli, Republic's legendary Chief Engineer, began studying how to incorporate a weapons bay into the swept-wing F-84F. Physically, the weapons bay was possible, but the added weight of structure around the fuselage cutout and the addition of racks, doors, and other equipment would have a serious effect on performance unless a new engine could be found. One possible choice was the General Electric J73 that was to be used in the North American F-86H Sabre. It was generally compatible with the F-84F airframe and was already scheduled for testing in a modified F-84F, designated YF-84J. However, the J73 was having development problems, and in any case would be in relatively short supply since the F-86 program enjoyed a high priority. Other possible choices were narrowed to the General Motors/Allison J71, in production for the Douglas B-66 Destroyer, and a version of the British Bristol Olympus, licensed for production by Wright and designated J67. Meanwhile, Pratt & Whitney was developing the J57, the engine that would power the North American F-100 Super Sabre, McDonnell F-101 Voodoo, and Convair F-102 Delta Dagger. Republic started investigating the installation of the J57 in the F-84F, along with a new wing and the internal weapons bay. This design was called F-84X by Republic.[4]

At the same time, work was started on a new design using the J73, known internally as the AP-63-FBX, although an entirely new aircraft, the AP-63 bore a remarkable resemblance to the RF-84F Thunderflash. The design featured an evolutionary version of the RF-84F wing fitted with outboard leading-edge flaps and conventional large-span ailerons. The all-flying stabilator was mounted midway up a sharply swept vertical stabilizer. The internal bomb bay could carry two 1,000-pound bombs, or a single "special" (nuclear) weapon weighing up to 3,400-pounds. Four T-130 0.60-caliber machine guns were mounted in the wing roots. With a combat weight of 27,550 pounds, the AP-63-FBX was projected to have a maximum speed of over 800 knots at 35,000 feet, making the aircraft comparable to the F-100.[5]

Both the F-84X and AP-63-FBX proposals were presented to the Air Force during February 1952, but at some point just prior to submittal, the AP-63's J73 was replaced with the Allison J71. Being an entirely new design, the AP-63 had a clear performance advantage over the modified F-84, with an associated increase in development time, cost, and risk. Nevertheless, in May 1952 the Air Force's Aircraft and Weapon Board recommended pursuing development of the new AP-63 instead of yet another modification of the venerable F-84. The aircraft was assigned the F-105 designation; however, no general operating requirements (GOR) existed for a new aircraft at the time.[6]

On 25 September 1952, Republic received a $13,000,000 letter contract [AF-33(600)-22512] for the development and production of 199 F-105As with a scheduled initial operational capability (IOC) of early 1955. The letter contract covered preproduction engineering, tooling design and fabrication, and materiel procurement. The F-105 was one of the first Air Force aircraft to be developed using the "weapons system" concept. Under the designation WS-306, Republic became the prime contractor responsible for the entire development and integration effort, including airframe, armament, engine, and all subsystems. The first public announcement of the program by the Air Force stated that Republic was developing a sophisticated, supersonic fighter-bomber capable of carrying nuclear weapons and powered by a General Motors/Allison J71-A-7 with a top speed of Mach 1.5. Precious few details were provided. But on 20 March 1953 Amendment No. 1 to the letter contract unexpectedly cut the program back to 37 F-105As to be delivered at a rate of 3 per month beginning in April 1955.[7]

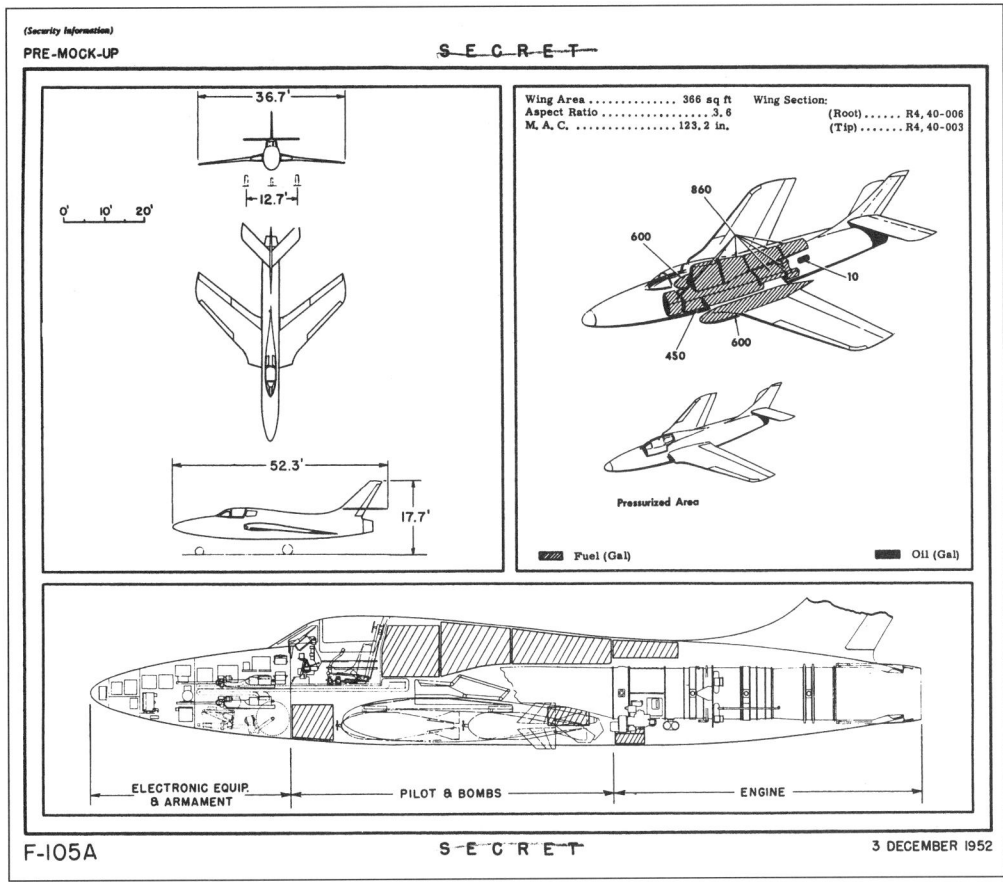

The AP-63-FBX design was submitted to the Air Force in February 1952. The Air Force approved the project in May 1952, and this was the general configuration of the F-105A in December 1952. Note the weapons bay under the wing and the four T-130 0.60-caliber machine guns in the nose. Although an entirely new aircraft, the AP-63 bore a remarkable resemblance to the RF-84F Thunderflash. With a combat weight of 27,550 pounds, the AP-63-FBX was projected to have a maximum speed of over 800 knots at 35,000 feet. This configuration would change considerably before the first aircraft was actually built. *(U.S. Air Force)*

F-105A

By mid-1953, the F-105A still bore an obvious lineage back to the RF-84F but had evolved considerably since the original AP-63 proposal. The thin, midmounted wing used an NACA 65A airfoil and was sharply swept (45° measured at 25 percent chord), with elliptical engine air inlets mounted on the leading edge adjacent to the fuselage. The inlets were fixed, with no variable area features. The all-flying stabilator was moved to a position low on the aft fuselage to maximize effectiveness at high angles of attack. The single pilot was housed under a streamlined canopy that was semiblended into the top of the fuselage. The well-tapered nose ended in a sharp point, as opposed to the more rounded nose originally proposed. The F-105 had a set of "clover-leaf" speed brakes attached to the end of the exhaust nozzle that unfolded to stand at nearly right angles to the airflow during braking. A small ventral fin was fitted on the bottom of the rear fuselage to provide additional stability at high speeds.

The internal weapons bay just below the wings, although larger in dimension than the original proposal, could carry less ordnance. The new weapons bay held only a single 2,000-pound conventional bomb, a single 3,400-pound nuclear weapon, or a removable

350-gallon fuel tank. There were four underwing hard points, each capable of carrying 1,000 pounds. The four T-130 0.60-caliber machine guns were retained.

The subject of internal armament came up in August 1953 when General Albert Boyd, Commander of the Wright Air Development Center (WADC), began to question the development of the new T-130 machine gun. General Boyd pointed to the results of Project GUNVAL, which had shown that 20-mm weapons possessed several advantages over 0.60-caliber weapons. Canceling the development of the 0.60-caliber weapon would save a great deal of money and allow a common ammunition to be shared with SAC, which had largely standardized on 20-mm weapons for bomber defense. The General suggested replacing the T-130 with the new General Electric T-171 20-mm revolver cannon.[8]

Also in August 1953, the WADC began a feasibility study of using the F-105 as a night and bad-weather reconnaissance aircraft. The list of desired features included low-altitude night photographic equipment, high-definition radar with a moving target indicator, radar recording, semiautomatic navigation devices, and infrared target sensors. The WADC asked Republic to confirm the feasibility of incorporating these items into a modified F-105 airframe. Republic indicated that the first RF-105 could be available in August 1957, assuming a timely approval to proceed.[9]

When the definitive AF-33(600)-22512 contract was issued on 25 September 1953, it included funds for 37 F-105s and 9 RF-105s. It was already obvious that the J71 would not meet either the thrust or schedule requirements for the F-105, so plans were made to temporarily install the new P&W J57 until the J71 matured. However, the J57 was in short supply, with numerous seemingly vital aircraft projects demanding their share all at once. A mock-up was inspected on 27 October 1953, including the J57 engine, and both T-160 and T-171 guns. The inspection resulted in 152 requests for alterations, most of them minor, although several would have to wait for resolution until the mock-up board issued its final report in March 1954. Republic still expected to meet the 1955 IOC date, but in December 1953 the Air Force suspended WS-306A entirely due to development delays at Republic, unexpected increases in gross weight, and the lack of a suitable engine.[10]

The increase in gross weight and development delays were largely by-products of the Air Force continuing to ask for more elaborate equipment and subsystems to enhance the aircraft's strike capabilities. What had started out as a relatively simple fighter was rapidly becoming the most sophisticated aircraft of its day, incorporating several electronic systems that would not have been possible a few years earlier. In an effort to find a suitable engine, the Air Force approached Pratt to see if the experimental J75 could be fitted to the F-105. The 16,000 lbf J75 was destined to power the Convair F-106 Delta Dart, and Pratt was sure it could be adapted to the Republic aircraft.

In February 1954 the program was reinstated with $49,900,000 of FY54 funding for 15 F-105As fitted with J57 engines, including one reconnaissance version. The switch to a less powerful engine for the YF-105As was an attempt to get the initial aircraft into the flight test program as early as possible. In case the J75 was delayed excessively, plans were also made to use the Wright J67 in production F-105s. During the same month the Air Force inspected the nose mock-up of the reconnaissance version with its KS-24A camera installation and found little wrong. In March the mock-up board finally issued their report on the inspection the previous October. Recommendations included adding a braking parachute, confirming the J57 as the interim engine and the J75 as the permanent engine, and installing a single General Electric T-171E-2[11] 20-mm cannon with 500 rounds of ammunition in the nose.

On 6 August the plan was modified yet another time. Ten of the 15 F-105s would be equipped with the J57-P-25, and the remaining five would be equipped with YJ75-P-1 engines. Procurement of the RF-105 was canceled except for some long-lead items and continued engineering studies.

Then, on 2 September 1954, Air Force Headquarters cut the program back to three aircraft, all powered by the J57. This was prompted by a review that concluded "... a general lack of confidence in this contractor fulfilling his commitments and meeting the performance guarantees on the F-105 weapon system." The order allowed local Air Force management to authorize the construction of "two or three additional aircraft if this is considered desirable." The actual stop-work order was issued on 9 September, limiting production to two J57-powered F-105As and a single YJ75-powered F-105B. Three weeks later this was amended to include three additional F-105Bs, and on 26 October a static test article was also authorized.[12]

On 1 December 1954, 31 months after the AP-63 design had received approval from the Air Staff, GOR-49 was finally issued. This GOR defined an aircraft having the performance and capabilities of the J75-powered F-105B, in essence making the F-105B an operational requirement. The requirements were largely based on the *Fighter-Bomber Weapons System's Military Characteristics* study published in 1951. GOR-49 itself would be amended four times during the first four months of 1955, and in the end specified that the F-105 should be powered by the J75, have an advanced, computerized, fire control system, an in-flight refueling capability, and an IOC of 1958—a three-year slip from the original date.[13] The F-105 was required to carry the T-171E-2 20-mm cannon, ninety-five 2.75-inch folding fin aircraft rockets (FFAR), a variety of tactical nuclear weapons, conventional bombs, and AIM-9 Sidewinder missiles.[14]

On 19 January 1955, the Air Force authorized the construction of two J57-powered YF-105A prototypes, ten YJ75-powered YF-105B[15] fighter-bombers, and three YRF-105B reconnaissance variants. What was different was how they were to be procured. Under the 1954 contract, the prototypes were to start a flight test program while final engineering was being completed on the production design. Any problems encountered during the initial flight test program would be factored back into the final design. But since a year had passed, and Republic had continued to work on the project without government funding, the production design was almost complete. Therefore all 15 aircraft would be procured under the Cook-Craigie concept of "concurrency."

The concurrency idea had originated during the late 1940s with Generals Orval R. Cook, Deputy Chief of Staff for Material, and Laurence C. Craigie, Deputy Chief of Staff for Development. The concept was to reduce the time needed between the start of a program and IOC. To achieve this, it was necessary to forgo the usual prototype stage and proceed directly to production. The first dozen or so examples off the production line would be dedicated to an accelerated flight test program, and any deficiencies found would be detected soon enough to be incorporated into production units intended for operational use. The same idea would surface again in the late 1960s and early 1970s when programs became so expensive that prototype efforts could not be afforded. Unfortunately, the idea has a major flaw—it assumes that any problems found will be minor in nature, clearly understood, and easily corrected. As would be discovered, this was not always the case.

On 25–26 January 1955 the J75 mock-up review was held at Pratt & Whitney. A total of 96 requests for alteration were made, but no major complaints were voiced. Ground testing had indicated that the engine would produce 15,500 lbf dry, and 23,500 lbf with afterburner. The engine appeared to be capable of speeds over Mach 2 and altitudes in excess of 70,000 feet. It appeared that the F-105 and F-106 finally had an engine.[16]

YF-105A

Not truly prototypes of the production configuration, the two YF-105As were powered by Pratt & Whitney J57-P-25 engines. Although readily identifiable as F-105s, they still had subsonic, elliptical engine inlets similar to those found on the RF-84F. The vertical stabilizer was considerably smaller than the production version, being lower in height, aspect ratio, and sweep angle. The fuselage of the prototypes was 2 feet shorter, 1 foot narrower, and 3 inches shallower than the later F-105B. The J57-P-25 developed 16,000 lbf at sea-level with afterburning. Its military rating was 10,200 lbf with a continuous thrust rating of 8,700 lbf. In accordance with the then-standard Air Force practice of carrying "buzz" numbers,[17] the F-105 was assigned FH, which was followed by the last three digits of the serial number, and was generally carried on the forward fuselage.

The decision was made to conduct the first flight from Edwards AFB, California, since it was considered safer, and more secure, than the Republic field on Long Island. Immediately after completion, the first YF-105A (54-0098) was disassembled for transport, Republic not even conducting an engine run-up test at the factory. The fuselage and miscellaneous parts were placed in one C-124 that departed Long Island for Edwards on 28 September 1955, while the wings, stabilators, engines, and more parts were placed in a second C-124 that departed the following day.

Crews from the Republic experimental shop accompanied the aircraft to Edwards to assist in assembly and preflight tests. On 22 October 1955 the first YF-105A made its maiden flight with Republic chief test pilot Russell "Rusty" M. Roth at the controls. A brief Republic press release stated that the aircraft had exceeded Mach 1 on that flight, but given the aerodynamics and power available, it could not have been by much. By the end of November the aircraft had made 12 flights and had been formally accepted by the Air Force. On 16 December 1955, the first YF-105A was performing a series of high-speed tests to evaluate maneuvering stability, including straight and level dashes followed by high-g turns and a complete roll. After completing several successful runs, the aircraft was traveling approximately 530 knots in a 5.5-g turn when the right main landing gear suddenly extended and was promptly torn off. Roth managed to recover and made a hard landing at Edwards, breaking the back of the aircraft in the process. The first YF-105A had logged approximately 22 hours during 29 flights up to that point and was returned to the factory for repairs. It never returned to flight status, and its exact fate remains unknown.[18]

The second YF-105A (54-0099) was also transported from Long Island to Edwards inside two C-124s, departing from the Republic plant on 8 and 9 December 1955. The aircraft made its maiden flight on 28 January 1956 with Rusty Roth again at the controls. The differences between the two aircraft were limited to the second having a smooth, unpetaled exhaust, while the first had the four-petal speedbrake system that became standard on production F-105s. The aircraft participated for a short time in the flight test program, but no record as to its fate has surfaced.

In the months following the first flight of the YF-105A, Republic's engineering department had more than 300 engineers working under the leadership of Sidney R. Huey. At the time, due to a shortage of office space at the Farmingdale factory, the engineers were assigned offices in the Dun & Bradstreet Building in New York City. Early in 1956 Republic renovated some office space above the factory floor, and all the engineering staff moved there in order to be closer to the production line. By this time it was obvious that the prototypes were in no way representative of the production aircraft. The YF-105A had incorporated all known aerodynamic features desirable for an aircraft designed to fly in the transonic and supersonic regimes. Its wings and tail surfaces were swept at sharp angles, and the wing was thin with leading and trailing edge devices to increase lift when desired. The all-moving horizontal stabilizer (called a stabilator) was chosen for increased effectiveness in transonic and supersonic flight, and was mounted low to aid in preventing violent pitch-up. The high wing loading and small wing area favored low drag at supersonic speeds. But in spite of all these features, the YF-105As were not truly capable of supersonic flight. The engine inlet design, heavily borrowed from the RF-84F, generated unwanted shock waves and did not allow sufficient airflow to the engine. Perhaps most of all, how-

The end of a very short flight test career. The first YF-105A (54-0098) made its maiden flight on 22 October 1955. On 16 December 1955, the aircraft was traveling approximately 530 knots in a 5.5-g turn when the right main landing gear suddenly extended and was promptly torn off. Republic chief test pilot Russell "Rusty" M. Roth managed to recover and made a hard belly landing at Edwards, breaking the back of the aircraft in the process. The first YF-105A had logged approximately 22 hours during 29 flights. *(Mike Machat Collection via Tony Landis)*

ever, the J57-powered aircraft had too much drag at transonic speeds, and as a consequence were significantly slower than expected.

Early in the design effort, the Air Force and Republic had requested engineers at the NACA Langley Aeronautical Laboratory to begin a test program in support of the F-105. Models were studied in a variety of wind tunnels and during free-spinning tests in a spin tunnel. NACA tested the airframe's basic aerodynamics, concentrating on any potential transonic and supersonic stability problems. Unquestionably the most dramatic influence of the NACA research was the application of the "area rule." This novel concept for drag reduction had been developed and verified by Richard Whitcomb and other researchers during December 1952 and had a profound effect on numerous supersonic aircraft designs. Whitcomb said that early wind tunnel data for the F-105 led to the same general conclusion as had data on the F-102 interceptor—the aircraft would not be able to comfortably exceed Mach 1. The answer lay in changing the contours of the fuselage to allow smoother pressure distribution, hence reducing transonic drag.

The tests also showed the aircraft shared the same directional stability deficiency of most other early supersonic aircraft, especially the F-100. The standard remedy was to increase the height of the vertical stabilizer, with NACA recommending a 32-percent increase plus the addition of a larger ventral fin to supply stability at very high angles of attack where the normal flow over the vertical might be partially blanketed by the down-

Although the two YF-105As differed considerably from the eventual production F-105s, the basic shape was evident. The second YF-105A (54-0099) was airlifted from Long Island to Edwards AFB inside two C-124s and made its maiden flight on 28 January 1956 with Rusty Roth again at the controls. The differences between the two aircraft were limited to the second having a smooth, unpetaled exhaust, while the first had the four-petal speedbrake system that became standard on production F-105s. *(U.S. Air Force Historical Research Agency Collection)*

wash from the wing. The vertical stabilizer's chord was also increased, and the rudder was made larger in both height and width. Leroy Spearman, an NACA engineer associated with the testing, proposed a series of schemes that were tested, but not adopted, to improve the F-105's directional stability, including horizontal strakes on the forward fuselage, cruciform strakes on the afterbody, twin ventral fins, and a variety of folding or retractable ventral fins.

Langley modified their F-105 model to include the taller vertical stabilizer, a larger ventral fin, and an area-rule fuselage. The latter involved bulging the fuselage aft of the wing and pinching it around the wing, resulting in the classic coke-bottle shape. At the same time they lengthened the nose slightly to accommodate a radar set the Air Force wanted but wouldn't fit in the existing design. The distinctive inlets of the production F-105 were also added at this point. These inlets were apparently heavily influenced by studies done for the ventral inlet of Kartveli's XF-103 triple-sonic interceptor. The XF-103 inlet was severely swept forward, and the new F-105 inlets were essentially half of this inlet laid sideways. The results showed that the modified shape easily slid through the transonic zone and well into the supersonic, even at the power level provided by the J57. At Kartveli's insistence, the final shape of the F-105B retained only about 80 percent of the fuselage bulge tested by NACA, but this still provided adequate performance levels. Approximately 5,000 hours of wind tunnel time had been provided by NACA for the tests. All the subsonic testing was done in the Langley 19-foot tunnel with a ¼th scale model. A ½₂nd scale model was

Many published reports indicate that the YF-105As did not have the ventral fin that would be standard on the production Thunderchiefs, but every available photo shows both YF-105As with a ventral fin. The most noticeable evidence that this is a YF-105A (54-0099) is the small vertical stabilizer. Wind tunnel testing would result in the height and chord of the vertical stabilizer being increased to provide directional stability at high speeds. *(Mike Machat Collection via Tony Landis)*

used for tests in the 8-foot transonic pressure tunnel, and also for tests in a 4-foot supersonic tunnel.

A major concern of Republic engineers early in the program was the possibility of high-speed flutter on the stabilator. In an attempt to find an answer, dynamically similar surfaces were instrumented, fitted to several sounding rockets, and launched from the NACA Wallops Island facility during 1953. The results were inconclusive. However, Langley had a 26-inch blow-down tunnel that could perform flutter studies, and on 18 October 1954 two models of the F-105A empennage were delivered for testing. These tests showed a major problem, and on 27 December one of the models tore apart during a test that simulated the normal operating speed of the F-105A. A few days later, the second model was also destroyed during a transonic test. The engineer in charge of the tests was Larry Loftin, who recalled that the exact nature of the problem was never understood, and the solution that was selected—simply overengineering the part for greater strength—was mostly of a brute-force approach.

During the same time period, concern about flutter affecting the vertical stabilizer prompted a similar series of tests. By late September 1955, tests showed that a serious buzz developed at some speeds and flight attitudes and was serious enough to cause the loss of the rudder on one model. Investigation showed that the rudder developed a torsional vibration that eventually amplified and destroyed the control surface. The eventual solution was to place a viscous damper at the hinge. During 1956 and early 1957, similar tests were run on the F-105B tail surfaces, which were designed for a higher speed range and used a different

The YF-105A demonstrated it could perform as a "buddy tanker" using a special centerline store equipped with a drogue refueling hose. This capability was incorporated into block-15 and later F-105Bs but was later deactivated. Here an F-104A prepares to take on fuel from the second YF-105A (54-0099) over Edwards AFB. *(Mike Machat Collection via Tony Landis)*

The two YF-105As used a straight air intake that was reminiscent of the RF-84F's. This intake was not optimized for supersonic flight and would be replaced by the forward-swept intake used on all production Thunderchiefs. One somewhat unwelcome change introduced on production aircraft was the elimination of the small windows behind the canopy. These provided a small amount of rearward vision, something production aircraft lacked. *(Mike Machat Collection via Tony Landis)*

style of connecting yoke between the two halves of the all-flying stabilator. Since these had the benefit of the earlier program's results, no serious problems were discovered.

North American YF-107A

The Thunderchief was not the only fighter-bomber on the drawing boards. On 4 March 1952, North American Aviation began an internal engineering study of a possible follow-on to the F-100A. The F-100B retained the original swept-wing planform of the F-100A but had a thinner wing cross-section with a 5 percent thickness/chord ratio rather than the 7 percent of the F-100A. The design used a 16,000 lbf J57-P-25 engine with a variable-area inlet duct and convergent-divergent exhaust nozzle. Dual main landing-gear wheels would make operations from unprepared airfields possible. The area-rule fuselage had an increased fineness ratio over the F-100A, and all the internal fuel was carried in integral wing tanks, with no provisions being made for carrying external fuel tanks. A maximum speed of Mach 1.80 at high altitude was anticipated, and production was expected to begin in 1955.

At the same time, North American began to study the feasibility of adapting the Super Sabre as an all-weather interceptor. The project became known as the F-100I (I for interceptor) or F-100BI, although these were not official Air Force designations.[19] This aircraft was similar in overall configuration to the F-100B except for a modified cockpit and a nose radome. In order to accommodate the radome, the variable-area air intake was mounted under the nose, similar to the later F-16. Provisions were made for underwing drop tanks, and an all-rocket armament was to be fitted. The F-100BI was intended to bear much the same relation to the F-100A as the F-86D did to the F-86A. On 20 October 1953 the factory designation NA-212 was assigned to the project, and work began on wind-tunnel studies and a detailed mock-up.

For unexplained reasons, in November 1953 North American began adapting the NA-212 as a fighter-bomber. Six hardpoints were added underneath the wings, and the wing structure, controls, and cockpit were revised accordingly. A single-point refueling capability was added, and the windshield and canopy were revised to improve the pilot's view. A retractable tail skid was installed and the flight control system was upgraded by the addition of pitch and yaw dampers.

Neither the F-100B nor the F-100BI attracted much interest from Air Force. Consequently, on 15 January 1954 the program was cut back drastically by North American president Lee Atwood. Plans to undertake full development were abandoned and the program was scaled back to a comprehensive engineering study.

On 16 April North American baselined the general configuration of the NA-212 as that of the F-100BI interceptor. However, later that month, North American learned that the Air Force was more interested in the fighter-bomber configuration. On 16 May 1954, work on the interceptor project was terminated and all efforts were concentrated on the fighter-bomber adaptation.

Among the changes needed to adapt the NA-212 as a fighter-bomber was the change from a 7.33-g to an 8.67-g load factor and the installation of a maneuvering autopilot, AN/APW-11A radar beacon, low-altitude bombing system (LABS), AN/ALE-2 chaff dispenser, AN/APS-54 radar warning system, and a cockpit computer. Larger and heavier wheels and brakes were specified, and provisions were made for electric fusing of external stores.

On 11 June 1954, the Air Force issued contract AF-33(600)-27787 for 33 F-100B fighter-bombers. On 8 July the Air Force notified North American that the aircraft's designation had been officially changed to F-107A. On 29 February 1956 the order was cut back to three prototypes and nine service test aircraft.

When Pratt & Whitney announced the development of the J75, North American modified the F-107 to accommodate the new powerplant. North American engineers also redesigned the vertical stabilizer into a single-piece, all-moving slab, similar to the one adopted for the North American A3J (later A-5) Vigilante carrier-based strategic bomber. A complex spoiler-slot-deflector system on the wings provided lateral control. One of the more novel features of the aircraft was the use of a side-stick controller instead of a conventional center-mounted control stick. This was an early attempt to provide a more ergonomic cockpit for the pilot.

Instead of an internal weapons bay, the F-107A carried a nuclear weapon with an integral 250-gallon fuel tank semirecessed under the fuselage. Unfortunately, wind tunnel tests showed that there would be major problems with weapon release and separation caused by airflow interference from the nose radome and chin air intake. In order to correct this problem, the air intake was moved to the top of the fuselage just behind the cockpit. Four M39A1 20-mm cannon were carried in the fuselage, while the wings provided four hardpoints for external stores. Production aircraft would carry an MA-12 fire control system that was generally similar to the MA-8 being developed for the F-105 but incorporating a different sight and a more advanced North American Autonetics R-14 radar.

The first YF-107A (55-5118) made its maiden flight on 10 September 1956 at Edwards AFB with North American test pilot Bob Baker at the controls. It went supersonic on its first flight, although there was some minor damage upon landing when the drag chute malfunctioned and the aircraft overran the end of the concrete runway and ended up in a ditch. The aircraft was quickly repaired and flew again three days later.

The North American YF-107A was also an evolutionary design, being based on the F-100 Super Sabre. The YF-107A introduced several advanced concepts, including an all-moving vertical stabilizer and a side-stick controller for the pilot. *(North American Aviation via the San Diego Aerospace Museum Collection)*

The YF-107A did not include an internal weapons bay like the F-105 design. Instead, weapons and fuel were carried semisubmerged in the bottom of the fuselage. The third aircraft carried small YF-107A markings on the extreme forward fuselage, unlike the first aircraft that had much larger markings behind the cockpit. *(North American Aviation via the San Diego Aerospace Museum Collection)*

A variable-geometry intake was installed on the third prototype (55-5120) but did not live up to its expectations. Most seriously, there was a buzz at high speeds, which was traced to instability of the airflow at the inlet. Several attempts to fix this problem failed and a final solution was never found.

The YF-107A was 61 feet 8 inches long, and the vertical stabilizer was 19 feet 6 inches high. The wing spanned 36 feet 7 inches and provided 376 square feet of area. The prototypes had an empty weight of 25,144 pounds and an estimated combat weight of 30,272 pounds. Demonstrated top speed was 1,295 mph at 36,000 feet. The maximum initial rate-of-climb was just under 40,000 feet per minute, and a 53,000-foot service ceiling was predicted.

The F-107A found itself in direct competition with the Republic F-105 Thunderchief for production orders. The F-105 was experiencing problems during its flight tests, in contrast to the F-107's, which were going relatively well. By mid-1956 the Air Force was debating canceling the F-105 and ordering the F-107 into production. But even under ideal circumstances the F-107 could not be ready as quickly as the F-105, and there was still the possibility that the F-107 itself would run into development problems. On 15 March 1957, the Air Force decided to stay with the F-105, and the F-107 was relegated to aerodynamic testing duties. The nine service test aircraft were canceled. Interestingly, the entire $105.8 million cost for the F-107 project had been paid for out of procurement support funds instead of the more normal RDT&E funds.[20]

The first and third YF-107As were eventually turned over to NACA for high-speed flight testing work. The first YF-107A reached NACA at the High-Speed Flight Station (HSFS) on 6 November 1957 and was assigned the NACA number 207. However, it was so mechanically

The placement of the air intake directly above and behind the cockpit resulted in an innovative, articulated canopy. Preliminary designs for the YF-107A used an intake mounted below the fuselage, much like the one used on the F-16. The North American Search and Ranging Radar (NASSR) for the YF-107 had been developed by the Autonetics Division of North American Aviation, and a modified version would later find use on the F-105D/F. *(North American Aviation via the San Diego Aerospace Museum Collection)*

unreliable that it was grounded by NACA after only four flights and was scavenged for spare parts to keep the other YF-107A flying.

The third YF-107A arrived at the HSFS in February 1958. Originally, NACA was interested in testing the variable-geometry intake, but this was cut short because of mechanical problems. Eventually, NACA gave up on the variable-geometry inlet altogether and it was bolted in a fixed position, limiting the aircraft's top speed to Mach 1.2. The YF-107As completed 40 test flights for NACA/NASA during 1958–59, and on the basis of this flight testing, North American refined the design of the side-stick planned for the X-15. The third YF-107A was damaged on 1 September 1959 when test pilot Scott Crossfield was forced to abort a takeoff because of control problems. Both tires blew and the left brake burst into flames. Crossfield was uninjured, but the resulting damage to the YF-107A was deemed to be too severe for economical repair and NASA decided to scrap the aircraft.

The other two F-107As still survive. After being retired by NASA, the first YF-107A was eventually acquired by the Pima Air and Space Museum in Tucson, Arizona, where it is now on display. The second YF-107A (55-5119) is in the Air Force Museum at Wright-Patterson AFB, Ohio.

Notes

1. A definition that has been said many times, but best by Larry Davis in *Republic F-105 Thunderchief*, Volume 18 in the WarBird Tech Series, Specialty Press, North Branch, MN, p 5. Used with permission.
2. AP stood for Advanced Project, an internal company design nomenclature.

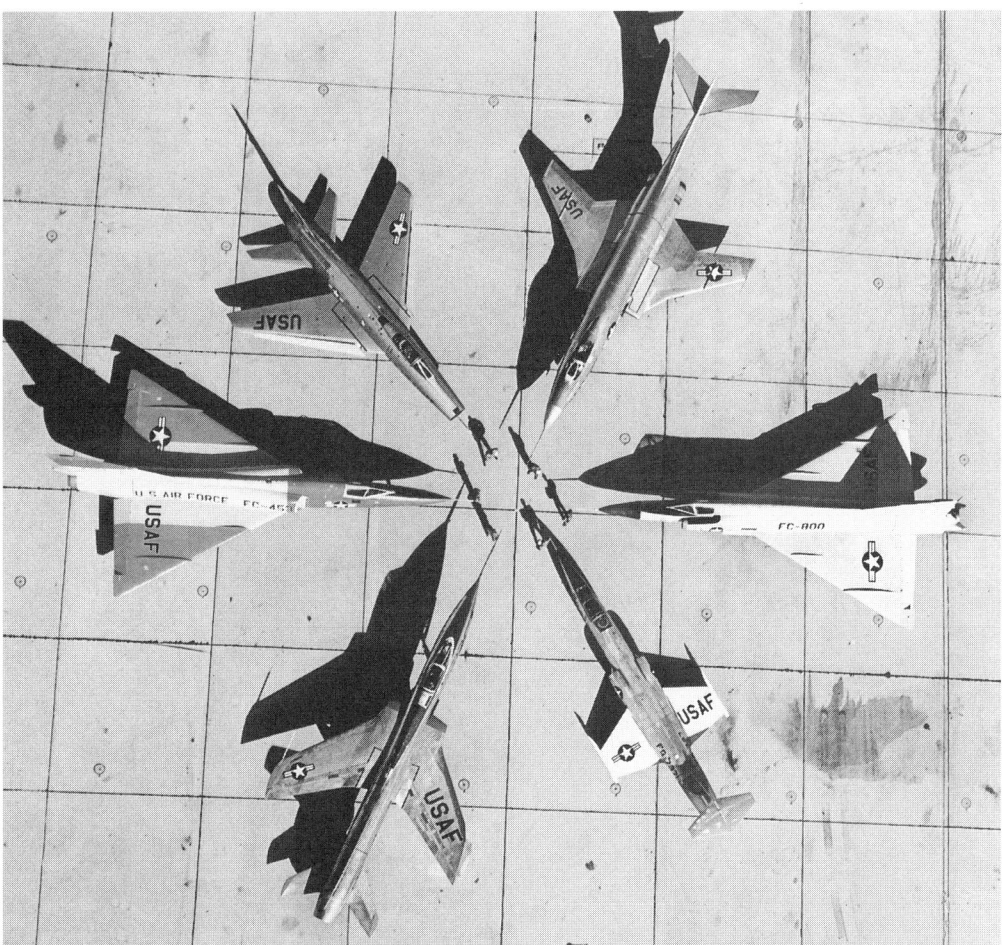

The North American F-100 Super Sabre (upper left) was the first operational supersonic fighter in the Air Force inventory, and each succeeding fighter would raise the performance level even further. The six fighters with 100-series designations became known as the "Century Series." Clockwise from the F-100A are the McDonnell F-101A Voodoo, Convair F-102A Delta Dagger, Lockheed F-104 Starfighter, Republic YF-105A Thunderchief, and Convair YF-106A Delta Dart. Note the distinctive straight intakes on the YF-105A. *(Mike Machat Collection via Tony Landis)*

3. Knaack, Marcell Size, *Post-World War II Fighters, 1945–1973*, Office of Air Force History, United States Air Force, 1986, p 329.
4. *Ibid.*, p 191.
5. Davis, Larry, and Menard, David, *Republic F-105 Thunderchief*, Volume 18 in the WarBird Tech Series, Specialty Press, North Branch, MN.
6. Knaack, Marcell Size, *Post-World War II Fighters, 1945–1973*, Office of Air Force History, United States Air Force, 1986, p 191.
7. *The F-105: A Chronology (1951–1973)*, History Office, Mobile Air Materiel Area, Brookley AFB, Alabama, September 1963, p 4.
8. *Ibid.*, p 6.
9. *Ibid.*, p 7.
10. *Ibid.*, pp 17–19.
11. The T-171E was a series of multibarrel cannons (subsequently designated M61), with the T-171E-2 using a linked ammunition feed system (where the expended shell casings are retained in a separate drum), and the T-171-3 using a linkless system (where the shell casings are returned to the same drum).
12. Knaack, Marcell Size, *Post-World War II Fighters, 1945–1973*, Office of Air Force History, United States Air Force, 1986, p 192.
13. *Ibid.*, p 192.
14. Designated GAR-8 at the time. In reality the carriage of AIM-9s was not authorized by the Air Force until 18 February 1957, and was not incorporated on the production line until early 1958.

15. The use of the Y (service test) prefix was confusing. The 19 January order used it. A 16 February 1955 revision to the contract omitted it on the last 13 aircraft. A special contract change notification on 8 March added it back to all 15 aircraft. On 4 November 1955 it was removed again from the last 13 aircraft, leaving only the two YF-105As as service test aircraft.
16. *The F-105: A Chronology (1951–1973)*, History Office, Mobile Air Materiel Area, Brookley AFB, Alabama, September 1963, p 21.
17. So named, or so the story goes, so that people on the ground that had been "buzzed" by overeager fighter pilots could readily identify the aircraft and report it to officials.
18. *The F-105: A Chronology (1951–1973)*, History Office, Mobile Air Materiel Area, Brookley AFB, Alabama, September 1963, p 34.
19. "I" is seldom used in official designations to avoid confusion with "1."
20. Knaack, Marcell Size, *Post-World War II Fighters, 1945–1973,* Office of Air Force History, United States Air Force, 1986, p 329.

In Production at Last—The F-105B

Republic Specification ES-349 detailed the F-105B and incorporated many of the design changes made during the NACA test program. Changes included reshaping the fuselage in accordance with the area rule, forward-swept engine air intakes, a taller vertical stabilizer, and a larger rudder. In addition, a ram-air intake was fitted to the base of the vertical stabilizer to capture cooling air for the engine compartment. The aircraft was powered by the Pratt & Whitney J75, rated at 23,500 lbf with afterburning. The F-105B had an empty weight of 23,873 pounds, and its maximum useful load was 20,580 pounds, including fuel and a 1,678-pound Mk 7 special store.[21]

The first YF-105A had used an exhaust nozzle covered by a four-segment set of doors, referred to as petals, that were hinged at the forward edge. These brake petals could be opened in various configurations. During landing, only the two side petals opened. There were two reasons for this—the lower petal had a ground clearance problem during landings, and the upper petal interfered with the deployment of the drag chute and also diminished the authority of the rudder during rollout before the nose wheel had touched down. For dive bombing, when controlling speed was crucial, all four petals were opened. The second YF-105A had eliminated these petals for unspecified reasons, but they returned on the F-105B.

On 25 January 1956 the Air Force baselined its preliminary Aircraft Weapon System Phasing Program. On 10 February 1956 the Air Force issued letter contract AF-33(600)-32216 covering the production of 55 F-105Bs and 17 RF-105Bs, although on 5 March this was increased to 65 F-105Bs and 17 RF-105Bs. In response to a requirement from the Air Training Command, on 10 May the Air Force added five two-seat F-105Cs to the order. Ultimately the Air Force Wanted to procure over 800 F-105s to equip 11 wings:[22]

The differences between the second YF-105A (54-0099) and the third F-105B (54-0102) show up well here. Most noticeable is the larger vertical stabilizer, although the overall shape of the rear fuselage is also different in accordance with the area rule principle. The Tactical Air Command was very interested in buddy refueling, and testing continued for most of the F-105's early career. *(Mike Machat Collection via Tony Landis)*

The only F-105B-6-RE (54-0111) shows the revised canopy area that deleted the aft glazed panels. The four-petal speed brakes are partially opened. Many early F-105Bs carried Indian-head motifs on the vertical stabilizer or forward fuselage at various times. *(Mike Machat Collection via Tony Landis)*

# Wings	Type	Command	Location	IOC
1	F-105	TAC	Langley AFB, Virginia	FY59
2	F-105	TAC	Clovis AFB, New Mexico	FY59/60
1	RF-105	TAC	Shaw AFB, South Carolina	FY59
1	F-105	ATC	Luke AFB, Arizona	FY59
1	F-105	ATC	Laughlin AFB, Texas	FY59
1	F-105	USAFE	Wethersfield, United Kingdom	FY60
1	F-105	USAFE	RAF Bentwaters, United Kingdom	FY60
1	RF-105	USAFE	Spangdahlem, Germany	FY60
1	F-105	FEAF	Misawa AB, Japan	FY60
1	RF-105	FEAF	Kadena AB, Okinawa	FY60

On 14 March 1956 the first J75 was delivered to Republic. This engine was installed in the first F-105B-1-RE (54-0100), which was subsequently airlifted to Edwards AFB on 29 April 1956. The aircraft made a 64-minute maiden flight from Edwards on 26 May with Henry G. Beaird, Jr. at the controls. Upon returning for landing, Beaird discovered that the nose gear would not extend and was forced to make a "fast and flat" emergency landing on one of the dry lake beds. The aircraft was not seriously damaged during the landing, but a crane operator managed to drop the aircraft while retrieving it and cracked the fuselage at station 285. This necessitated major repairs, and the aircraft was out of service for several weeks.

On 19 June 1956 Republic officially asked the Air Force to approve the name Thunderchief for the F-105, and on 25 July the Air Force agreed. This continued a long-standing Republic tradition—Thunderbolt, Thunderjet, Thunderstreak, Thunderflash, and now Thunderchief.

Republic spent July and August 1956 answering a series of questions regarding the cost of the F-105. The initial F-105Bs were costing $2,090,770 per aircraft, and the government needed to know how much later production examples would cost. Republic responded that—assuming a two aircraft per day manufacturing rate—the 400th F-105 would cost $536,000 and the 850th aircraft would run $445,000. The cost of the five C-models on order was estimated at $3,733,510 each, a significant premium over a standard B-model. By October 1956 the projected F-105 production had stabilized at 347 F-105Bs that cost $697,845 each, and 76 two-seaters costing $805,490 each, not including engines and weapon system components.[23]

The second F-105B (54-0101) was airlifted to Edwards AFB on 8 December 1956 and made its first flight on 28 December. On 19 January 1957 the aircraft lost its right main wheel during landing, sustaining minor damage that was quickly repaired. Then, on 30 January 1957, the main gear failed to extend, resulting in a belly landing that caused significant damage to the lower fuselage. In the next several weeks, this would also happen to the fourth aircraft. Concerned that this was becoming a trend, Republic engineers started looking for answers. What they found was that an air-bypass duct that led from the air intakes to the wheel wells created enough suction to hold the landing gear in the up position at high engine rpm and low airspeed. A change to the ducting cured the problem.

The first Thunderchief to make its maiden flight from Farmingdale was the third F-105B (54-0102), which did so on 29 April 1957. The AN/APN-105 Doppler navigation system was approved for use on the F-105 during July 1957, and the Air Force described this system as "the newest, lightest, smallest, and most accurate navigation system available . . ." The sys-

A great deal of secrecy surrounded the early F-105 aircraft, but the Air Force finally began to allow them to be put on display at air shows. They were an instant hit. The sheer size of the single-seat fighter excited most onlookers, while the shape was symbolic of the streamlined vehicles of the 1950s. Here the first F-105B (54-0100) is the subject of interest to some young spectators. *(Mike Machat Collection via Tony Landis)*

tem did not require a ground station and could operate without the benefit of periodic radar updates during periods when the radar was either inoperative or switched off to avoid detection. The APN-105 furnished the pilot with present position, course, distance to destination, ground speed, ground track, and drift angle data. The system would be installed on block-15 aircraft and retrofitted to block-10.[24]

Up until this time, the Air Force had remained silent about the F-105 project, and only a single, retouched photo of the first prototype had been released to the public, along with a few very cryptic press releases. This did not stop the French magazine *Aviation* from publishing a major article on the aircraft on 1 May 1957, causing a great deal of commotion within Republic and the Air Force. Since the secrecy surrounding the project had been compromised, the Air Force allowed Republic to publicly display an F-105B for the first time at the 28 July 1957 Andrews AFB, Maryland, air show. Interestingly, by this time the F-105B was already considered obsolete, and most Air Force attention had switched to the proposed "all-weather" F-105 models using the new R-14 North American Search and Ranging Radar System (NASARR). These were referred to as either F-105B/C(AW) or F-105D/Es, depending on which part of the Air Force was talking.[25]

On 14 October 1957, an F-105B (54-0107) began a series of environmental tests in the climatic laboratory at Eglin AFB. The aircraft, with a complete XMA-8 fire control system, was subjected to a variety of extreme cold (–65°) and hot (+100°) temperatures, ice, rain, wind, and other climatic extremes. The test concluded on 19 November 1957 with no major problems identified. This testing was actually completed four months ahead of schedule and

The first six F-105Bs were relegated to test duties for their entire Air Force careers. This is the third F-105B-1-RE (54-0102) with orange markings on the vertical stabilizer. Note the long test instrumentation boom protruding from the nose. *(Mike Machat Collection via Tony Landis)*

provided valuable data prior to dispatching the ninth aircraft (54-0107) to Alaska for an operational cold weather evaluation.[26]

On 22 November 1957 the Air Force issued GOR-49-1, which consolidated all F-105 requirements into a single document that would be applicable to the "mass produced" aircraft.[27] This finalized the installation of the APN-105 Doppler navigation system, General Electric MA-8 fire control system, and added the capability to carry different nuclear weapons (Mk 28, 43, 57, and 61) and a possible underwing tow-target system. The MA-8 was one of the first truly integrated fire control systems designed for a tactical aircraft. The new requirements also specified that production aircraft should be equipped to carry at least two[28] of the air-to-ground missiles specified in GOR-166, which had been issued in October 1957.[29]

Republic had built the initial batch of 15 aircraft under the Cook-Craigie plan with the expectation that all of them would be used in the flight test program. Of the 15, two were the YF-105As, four were F-105B-1-REs, two were similar[30] JF-105B-1-REs, a single JF-105B-2-RE, five F-105B-5-REs, and a single F-105B-6-RE. Of these, the ninth aircraft was at Eielson AFB, Alaska, flying cold weather operational tests as part of Project Raw Deal, the three JF-105s were at Farmingdale conducting systems tests, the four block-1 aircraft were at Edwards, one block-5 was at Eglin AFB testing the XMA-8, a nonflying example was at Wright-Patterson AFB undergoing static-load tests, and three aircraft (54-0109/0111) were assigned to a joint Republic/Air Force operational combat evaluation program. The Alaskan aircraft would later be flown at Wright-Patterson during all-weather evaluations, and then at El Centro AFB and Vincent AFB (outside Yuma, Arizona) for hot weather evaluations.

The Air Force Flight Test Center at Edwards published a preliminary flight test evaluation on 9 June 1957. The tests had been conducted between 8 January and 7 March 1957 and consisted of 18 flights totaling 13 hours and 45 minutes. According to the report, the F-105B ". . . has the potential of becoming an excellent fighter-bomber. But it needs a large number

Between 14 October and 19 November 1957, an F-105B (54-0107) began a series of environmental tests in the climatic laboratory at Eglin AFB. The aircraft was subjected to a variety of extreme cold (−65°) and hot (+100°) temperatures, ice, rain, wind, and other climatic extremes. The chamber allows the engine to be run at moderate power levels while the aircraft is inside. Note the frost covering almost every surface. *(U.S. Air Force Historical Research Agency Collection)*

of improvements before it could be considered acceptable for operational use." The report disclosed that the F-105 "is capable of 1.95 Mach number in level flight at 35,000 feet and 1.49 Mach number at 20,000 feet under standard day conditions." However, the report went on to say that Mach numbers in excess of 1.8 were "not considered usable because of poor acceleration characteristics." This was borne out by the fact that the F-105B required nearly nine minutes to accelerate from Mach 1 to Mach 1.95 at 35,000 feet. Noteworthy was the fact

A nonflying structural test article was used in Building 65 at Wright Field to determine the integrity of the F-105 airframe. This testing uncovered major structural problems with the early F-105Bs, but not before several aircraft had been lost in accidents. A structural splice plate at the top centerline of the fuselage was prone to failure, and a redesigned splice plate was installed as part of Project Back Bone on all B-models and early D-models that had used the same construction technique. *(U.S. Air Force Museum)*

that fuel reserves were nearly depleted by the time the aircraft reached its maximum speed. The tests concluded that the aircraft had a usable combat ceiling of 46,500 feet, but could be zoom-climbed to 63,000 feet with a significant loss in speed.[31]

With the planned IOC date already three years behind schedule, TAC elected to conduct the Category II operational test and evaluation (OT&E) with one of the units due to receive the fighter. It was argued that this would save considerable time between testing and operational service. This testing began on 8 January 1957 and was scheduled to end on 30 November 1959, although in fact it extended until 30 March 1960.[32]

During the F-105's gestation period, there had been a great deal of negotiation between the Air Force development commands and the user commands on exactly what weapons the F-105 would carry when it entered service. On 28 August 1957 the matter was finally settled. The list of weapons to be carried externally included MC-1 750-pound massive toxic bombs, smoke/chemical dispensing tanks, M-38 (Y-28E2) fragmentation cluster bombs, Mk 83 1,000-pound low-drag, general-purpose bombs, M-117 750-pound general-purpose bombs, LAU-3/A 19-round rocket launchers, LAU-2/A 30-round rocket launchers, AIM-9 Sidewinder[33] air-to-air missiles, Navy Aero X5A flare dispensers, ALE-2 chaff pods, "universal" ECM pods (later designated AN/ALQ-31), and MA-2 rocket training launchers. Internal weapons remained the TX-38 and TX-43 nuclear bombs, the MN-1 practice bomb dispenser, and the 20-mm cannon. External fuel tanks could be carried on the inner wing stations and the fuselage centerline, and the weapons bay could accommodate an internal fuel tank.[34]

An obviously posed photograph of the weapons that could be carried by the F-105B. The covered shapes marked "classified" were nuclear weapons. Note the buddy refueling pod to the left of the fuselage. The F-105B would never have to carry weapons into combat. *(Mike Machat Collection via Tony Landis)*

There was an interesting proposal from MIT to use the F-105 as the launch platform for the Republic AP-95 air-to-surface missile. According to Republic and MIT, this missile would give the F-105 the capability to strike targets from a distance of 500 nautical miles. No further information could be located on this project.[35]

By 15 April 1958, 643 flight tests had been completed, including 163 by 20 different Air Force pilots, with 7 different Republic pilots accumulating the remaining 480 flights. Republic had 1,300 engineers working on the project and had produced over 10,000 engineering drawings. A press release noted that there were over 65,000 individual components

It is unusual to see the weapons bay fuel tank in its lowered position. The weapons bay doors retracted inside the fuselage, minimizing drag during high-speed weapons deployment. This F-105D-15-RE (57-5791) still wears its natural metal finish. *(Cradle of Aviation Museum Collection via Ken Neubeck)*

on the aircraft and that over 5 million man-hours of engineering had been used to develop it. In late April, during a practice fly-by at Eglin AFB in preparation for a fire-power demonstration, the second F-105B (54-0101) developed engine trouble and crashed. Fortunately the pilot ejected successfully.

On 1 May 1958, the 335th Fighter (day) Squadron was formed within the 4th Fighter (day) Wing at Seymour-Johnson AFB. Shortly thereafter, they were redesignated 335th Tactical Fighter Squadron (TFS) and 4th Tactical Fighter Wing (TFW), respectively, conforming to the Air Force's new organizational policies. Detachment 1 at Eglin AFB was formed to participate in Category II testing with members from the Weapons Systems Project Office, Air Materiel Command, Air Force Flight Test Center, Air Proving Ground Command, Wright Air Development Center, and Republic Aviation. The Project Officer was Lieutenant Colonel Robert R. Scott, commander of the 335th TFS. Category I and II flight testing was continuously interrupted or delayed because of various problems with the aircraft and its systems. In addition, special tests of the new weapon system were necessary, and a variety of problems required engineering changes, further delaying operational testing.

Category III testing was postponed until modifications to the fire control system could be completed. In late July 1960, the 335th and newly formed 334th TFS finally began Category III testing at Nellis AFB and Williams AFB, respectively. The testing was completed on 15 August after being hampered by a severe shortage of spare parts. During the test, the poor performance of the MA-8 fire control system confirmed that the recent modifications had not actually fixed the problems in either accuracy or reliability.[36]

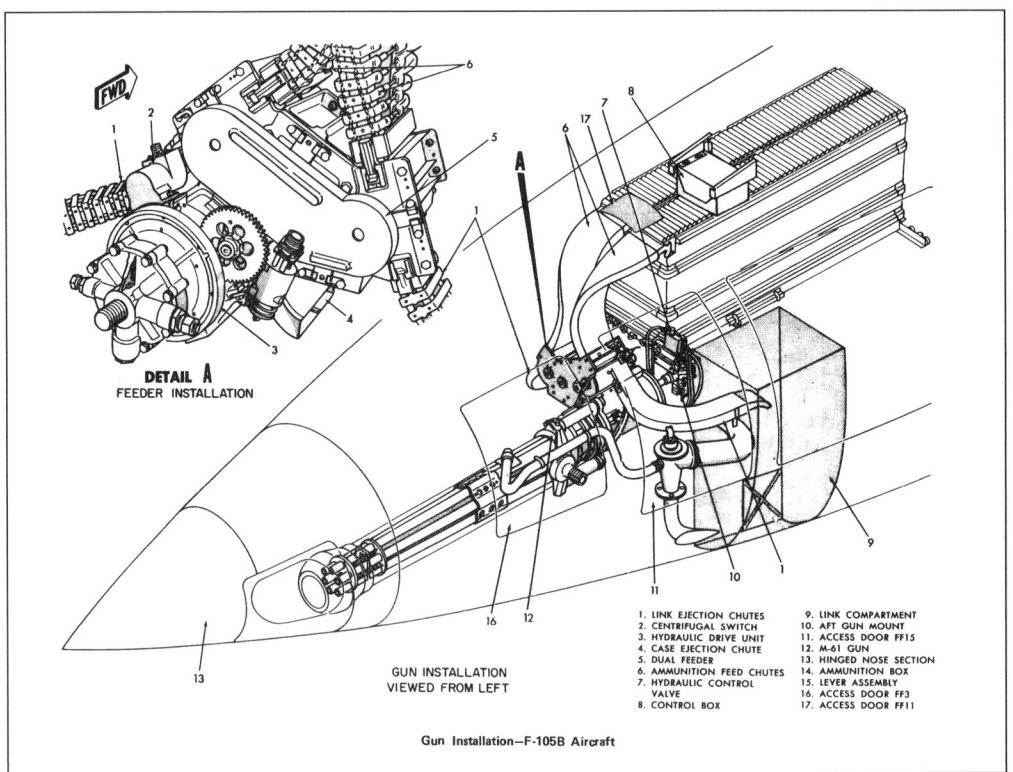

A major internal difference between the F-105B and later models was the configuration of the cannon and its ammunition feed. In the F-105B the M61A1 cannon used a linked ammunition belt that fed rounds into the breech, then deposited the spent casings into a separate container. This occupied a great deal of internal volume and would be replaced by a revised installation on the F-105D/F. *(U.S. Air Force)*

The last F-105B-1-RE (54-0103) showing the wear of a hectic flight test schedule. The red paint on the nose is flaking off, exposing the natural metal underneath it. The Air Force Flight Test Center markings on the vertical stabilizer were normal for the era. The size of the centerline external fuel tank was limited by its clearance with the ground, especially when the nose strut was compressed during landing. *(Mike Machat Collection via Tony Landis)*

The first F-105B-5-RE (54-0104) was externally representative of the production B-models. The only major exterior change made after this was the addition of a tail hook integrated into the ventral fin. It was normal for the lower-speed brake petal to droop as hydraulic pressure bled down. *(Mike Machat Collection via Tony Landis)*

The first production F-105B-10-RE (57-5776) was accepted by the Air Force at Farmingdale on 27 May 1958, and sent to the 335th in August. Production was running seriously behind schedule, and on 11 October 1958 the second production F-105B (57-5777) was lost during an acceptance flight. The pilot reported that he had felt a jolt during climb-out and received a landing gear warning light. Two F-86s in the area performed a visual inspection and reported that the right landing gear was hanging down, twisted, and swinging. The pilot elected to eject over Long Island Sound. With production slippage and mandatory modifications, it took 10 months before the 335th was up to strength with 18 aircraft. The four squadrons of the 4th TFW did not come up to authorized strength until 1960.[37]

The Thunderchief ejection system was tested at the Supersonic Military Aircraft Research Track (SMART) at Hurricane Mesa, Utah, in late June 1958. The test results showed an "almost" zero-altitude capability at 175 knots using the M-3 catapult, B-5 parachute, and MA-6 automatic lap belt. Officials believed that the inclusion of the D-ring lap belt under development would give a true zero-altitude capability. The tests also revealed that the escape system was not satisfactory at supersonic speed and low altitudes (Mach 1.2 at 5,000 feet), where the seat had a tendency to hit the vertical stabilizer. Subsequent modifications to the catapult solved the problem.[38]

Late in 1958 Republic took the unusual step of shutting down the F-105 production line. This was an attempt to build an inventory of critical parts from subcontractors and also to provide time to make Air Force mandated changes in aircraft already on the line. The Air Force sent an investigation team to figure out why the schedules were lagging and discovered several contributing causes. The Cook-Craigie plan did not work, particularly since

The forward-swept air intakes show up well here. The F-105 had a very small wing given its operating weight. Although this hindered maneuverability, it served the aircraft well in the fighter-bomber role. *(Mike Machat Collection via Tony Landis)*

the F-105 was going through so many changes so quickly that any problems found on early aircraft probably bore no significance to later aircraft. The Air Force itself shared some of the blame for the slippage, since they had specified many and continuing changes to the production configuration, resulting in seven distinct blocks of B models that totaled only 78 aircraft (including RF/JF-105Bs). In any case, the Air Force was critical of Republic's decision to halt the production line and suggested the Republic develop a better system for configuration management and problem reporting. The production line was restarted in mid-December, and that ended the crisis.

An example of the running changes included the way the canopy was attached. The four F-105B-1-REs (and both YF-105As) had canopies hinged at a single point at the rear of the top portion of the frame. This permitted the incorporation of a glazed panel aft of the canopy on each side, allowing the pilot at least some vision to his aft sides. These canopies were manually operated. Later F-105Bs introduced electrically operated canopies attached on each side of the rear frame, eliminating the rear glazed panels. This provided a slightly more robust hinge, but at the expense of essentially all rearward vision. The block-10 and block-15 aircraft that came off the production line were equipped with J75-P-5 engines instead of the earlier J75-P-3s. Block-10 introduced antiskid brakes and a cartridge-type starting system for the engine, increased the oxygen system capacity from 5 liters to 10 liters, and linked the rudder and stabilator to the automatic flight control system. Block-15 included provisions for a buddy[39] refueling system and added an integrated oxygen system and survival kit to the ejection seat. In block-20, the engine was upgraded to the more pow-

The business end of the J75 turbojet. Almost 10 years of teething problems with the J75 took their toll in F-105 accidents, but the engine provided a tremendous amount of thrust for the era. A pair of M118 3,000-pound bombs is being carried on the inboard pylons, while Mk 83 1,000-pound bombs are carried on the outer pylons. *(Mike Machat Collection via Tony Landis)*

erful J75-P-19, which provided another 1,000 lbf, and a chaff dispenser was added to the ejection seat. A more reliable electronics suite and antiskid brakes were also incorporated into block-20 aircraft.[40]

During the first few months with the 335th, less than 25 percent of the new aircraft were flyable at any one time. Additionally, the autopilot, the central air data computer, and the MA-8 fire-control system were unreliable, and there was a continuing shortage of spare parts. By mid-1959, the Air Force had completed 2,323 hours of flight testing during 2,158 flights. Between July and December 1959, aircraft of the 4th TFW accumulated 12,326 hours, all without a fatal accident. However, operational costs were high—the F-105B cost $718 per flight hour—compared to $538 for the F-100D, $611 for the F-102A, and $395 for the F-104C. Most of these costs were directly attributable to the fact that the F-105B was requiring 150 maintenance man-hours per flight hour, an abnormally high rate.

Barrier testing of an F-105B modified to include an arresting hook began at Edwards on 11 February 1959. Within 2 weeks 10 test runs had been conducted at various speeds against a chain barrier. Further testing was conducted using a "water-squeezer" barrier. The tests demonstrated that the arresting hook was satisfactory for emergency use, and the equipment would be included beginning with block-20 aircraft.[41]

In mid-1959 the Air Force ordered a group of improvements to all in-service F-105Bs, collectively known as Project Optimize. The modifications brought all aircraft up to block-20 standards, including retrofitting the antiskid brakes and the J75-P-19 engine. Modifications to the MA-8 fire control system, central air data computer, and autopilot to enhance

Some interesting weapons configurations were tested on the F-105. Here, an F-105B-20-RE (57-5836) is carrying two multiple ejector racks under the fuselage, plus another one on each inboard wing pylon. This allowed 26 Mk 82 500-pound bombs to be carried, but it was never approved for operational use. *(San Diego Aerospace Museum)*

their reliability and decrease maintenance requirements were also accomplished on all aircraft. At the same time, early B-models received the tail hook that had been introduced on block-20 aircraft. Project Optimize was supposed to last four months but ended up taking almost a year, finally being complete in July 1960. Most of the delay was attributed to a lack of spare parts and funding, although a certain amount of bureaucracy and uncertainty that the modifications would actually help anything also delayed completion.[42]

Despite its problems, the F-105B could be a performer. This was demonstrated on 11 December 1959 during Project Fast Wind. An F-105B, piloted by Brigadier General Joseph H. Moore (commander of the 4th TFW), set a world speed record of 1,216.48 mph over a 100-kilometer closed course. The course was 38,000 feet over Edwards AFB, carefully watched by observers on the ground, and tracked by radar to validate the record. The previous record had been set in June 1959 at 1,100.42 mph by a French Nord-Griffon II.[43]

Even with the improvements offered by Project Optimize, none of the 56 F-105Bs assigned to TAC were deemed operationally ready during the first three months of 1960. The unreliability of the MA-8, central air data computer, and autopilot remained the principal problems, although the J75 was also proving to be somewhat troublesome. During December 1961, all F-105s were grounded for a short period following the failure of a main forged-steel frame at fuselage station 442.785 during laboratory fatigue tests at Wright-Patterson. Further analysis indicated the frame retained some load-carrying capacity even after fracturing, and the grounding was lifted. Republic was able to carry out corrective work in the field without removing individual aircraft from service for a prolonged period.

This was the normal pre-Look Alike markings on the F-105B in natural metal finish. Look Alike would paint the entire aircraft with an aluminized silver acrylic-lacquer in an attempt to seal the seams. *(John Golden Collection via Ken Neubeck)*

When production of the F-105B terminated in December 1959, a total of 75 F-105Bs (and three JF-105Bs) had been accepted by the Air Force, including the various test aircraft. Three were accepted in FY57, 6 in FY58, 28 in FY59, and 38 in FY60. The total cost per aircraft was listed as $5,649,543, broken down as $4,914,016 for the airframe, $328,797 for the J75, $141,796 for avionics, and $264,934 for armament and ordnance. These costs were exclusive of $261,793 in modification expenses. This easily made the F-105 the most expensive of the early Century Series fighters.[44]

Only one Air Force wing was combat operational in the F-105B, the 4th TFW at Seymour-Johnson AFB. By 1960, the 4th TFW had three of its four squadrons equipped with F-105Bs, although it was always slightly below strength with only 60 F-105Bs available at any one time (a normal Air Force wing had 72 aircraft). The F-105 was so different in both its characteristics and mission that the Air Force had several strict requirements before a pilot could fly the aircraft. These included a minimum of 200 flight hours in other Century Series aircraft such as the F-100, completion of a 16-day ground school, and 75 hours flying time before he was considered qualified in the F-105B.

Two F-105Bs were involved in major accidents at Nellis during June 1962 and, on 20 June, all F-105s were grounded for inspection and modifications. The Air Force hurriedly modified 36 F-105Bs under Project Big Bear. The flight controls had to be rerigged and the hydraulic lines and electrical wiring, which had a tendency to chafe, had to be inspected and replaced as necessary. In addition, the fuel systems were modified to stop persistent leaks, and all outstanding change orders were incorporated. It took 858 man-hours to modify each aircraft. Similar modifications to the remaining F-105Bs were accomplished as part of Project Look Alike, with the last aircraft completed on 4 November 1962.[45]

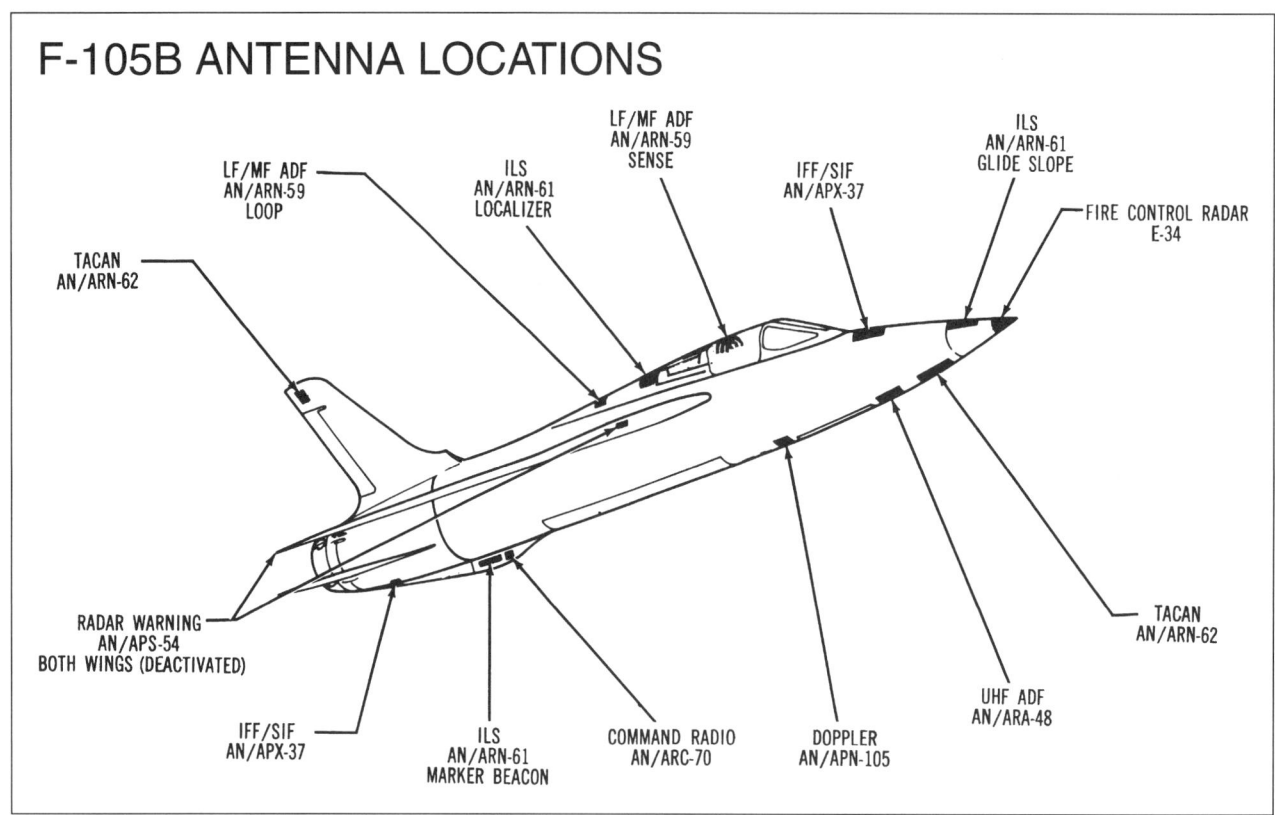

Although considered very complex when it was developed, the avionics suite on the F-105B was actually fairly simple, even by later F-105 standards. By the time this drawing had been released, the APS-54 radar warning system had already been deactivated. The F-105B would never be fitted with another radar warning system. *(U.S. Air Force)*

The first F-105B-20-RE (57-5803) used conventional instrumentation, unlike the later F-105D. The large handle at the upper left corner is for the drag chute, with the landing gear controls directly beneath it. The weapons controls are just to the right of the landing gear switches. The pylon jettison switches are located on the extreme right. *(Mick Roth)*

Three F-105Bs (57-5819, 5833, and 5835) from the 466th TFS (Air Force Reserve) at Hill AFB take on fuel from a KC-135. The F-105B was never fitted with a boom refueling receptacle, forcing the tankers to carry the drogue adapter on the boom. Although the B-model would never be committed to combat, it nevertheless received the same camouflage as the D/F-models. *(Mick Roth Collection)*

This was not the end of Project Look Alike, however. During the inspections, a serious corrosion problem was discovered, caused by moisture seeping in through seams in the skin. As an extension to Look Alike, all F-105s were painted in an aluminized silver acrylic-lacquer (1 gallon of clear lacquer mixed with 1 gallon of thinner and 12 ounces of aluminum paste)[46] in an attempt to better seal the aircraft and prevent water seeping into electronics areas. Teams of technicians from the Air Logistics Command, augmented by technicians and engineers from Republic, performed the modifications at the Mobile Air Material Command facilities at Brookley AFB, Alabama.

The F-105B almost went to war when President John F. Kennedy confronted the Soviet Union over nuclear missile installations in Cuba during mid-October 1962. On 18 October, the 4th TFW was ordered to deploy to McCoy AFB, Florida. There they sat one-hour "ready alerts" to conduct, of all things, air defense missions! For the remainder of the Cuban Missile Crisis, the 4th TFW flew air defense missions over southern Florida, returning to Seymour-Johnson on 29 November 1962. Interestingly, the 4th had to borrow 20 F-105Ds from the 4500th Combat Crew Training Wing at Nellis to bring the wing up to operational strength.

About two months after the end of the Cuban crisis, the 4th TFW began replacing their F-105Bs with new F-105Ds, and the F-105B was rather quickly phased out of the active inventory. The aircraft were reassigned to the Air National Guard (ANG), with the first reaching the 141st TFS of the 108th TFW, New Jersey ANG, on 16 April 1964. The B-model was so different from subsequent D/F-models that its value as a trainer was very limited, but nevertheless a few F-105Bs survived with the 23rd TFW at McConnell AFB, Kansas, until late 1969. The 466th TFS of the 508th TFG (Air Force Reserve) at Hill AFB, Utah, gave up its C-124s for F-105Bs in January 1973, receiving the aircraft previously flown by the New Jersey ANG, when they transitioned to the F-106. The 508th would finally turn in their Thunderchiefs in 1981, replacing them with F-16 Fighting Falcons.[47]

An F-105B-20-RE (57-5808) assigned to the 561st TFS lifts off the runway at NAS Miramar in August 1968. Most of the F-105Bs had been phased out of front-line service into National Guard squadrons by 1964, but the 23rd TFW still had several B-models for transition trainers in the late 1960s. Although the weapons systems were completely outdated, the F-105B's handling qualities were sufficiently close to the D/F-model to make viable trainers. *(Warren Bodie via the Larry Davis Collection)*

This F-105B-20-RE (57-5803) is assigned to the 466th TFS (Air Force Reserve) at Hill AFB. The size and color of the AFRES markings on the aft fuselage varied over time, with these representing the largest and most visible. *(Ben Knowles via the Mick Roth Collection)*

This F-105B-20-RE (57-5820) was photographed at George AFB in July 1980. At the very end of their careers, the F-105Bs of the 466th TFS (Air Force Reserve) at Hill AFB received wraparound camouflage schemes and black tail codes. Note the deployed in-flight refueling probe on the far side of the fuselage. *(Mick Roth)*

RF-105B/JF-105B

Three aircraft of the initial F-105B order for 15 were started as reconnaissance versions with a KS-24A camera system in a modified nose. These were given the Republic designation AP-71 (under RAC Specification ES-350), with the Air Force calling them RF-105B. The 20-mm cannon was removed to make room for cameras, so a pair of T-160 (M39A1) 20-mm cannons were installed in external blisters along the side of the fuselage.

During April 1955 a study was conducted on three additional RF-105 variants. These included night-all-weather, electronic reconnaissance, and weather reconnaissance versions. Although some of the capabilities appealed to TAC, both TAC and WADC were concerned that authorizing any of these new versions would upset the progress of the entire F-105 project by overdiluting Republic's engineering capacity. There were also concerns that the requirements being contemplated for the electronic version "took advantage of the farthest limits of electronic reconnaissance state-of-the-art" and appeared to require a "major break-through to unattended ferret operation."[48]

A mock-up of the RF-105B installation (which differed from the RF-105A installation that had been inspected in November 1954) was held on 1–2 June 1955. This review resulted in a recommendation for more cockpit reviews after Republic had installed instruments and cockpit lighting. Republic proposed deleting the 20-mm cannon mounted on the side of the fuselage, but the Air Force insisted they be retained. The camera installation included a forward oblique television camera, a oblique film camera, a trimetragon camera installation, and a vertical film camera. The cockpit included a large television screen that could show the signal from the oblique TV camera.

On 20 July 1956, the Air Force terminated all work on the RF-105 program. This was not the end of it, however. As early as October 1956, only three months later, the Air Force issued letter contract AF-33(600)-33874 directing Republic to build forward fuselage and cockpit mock-ups of the three variants discussed in April 1955. Work on the night all-

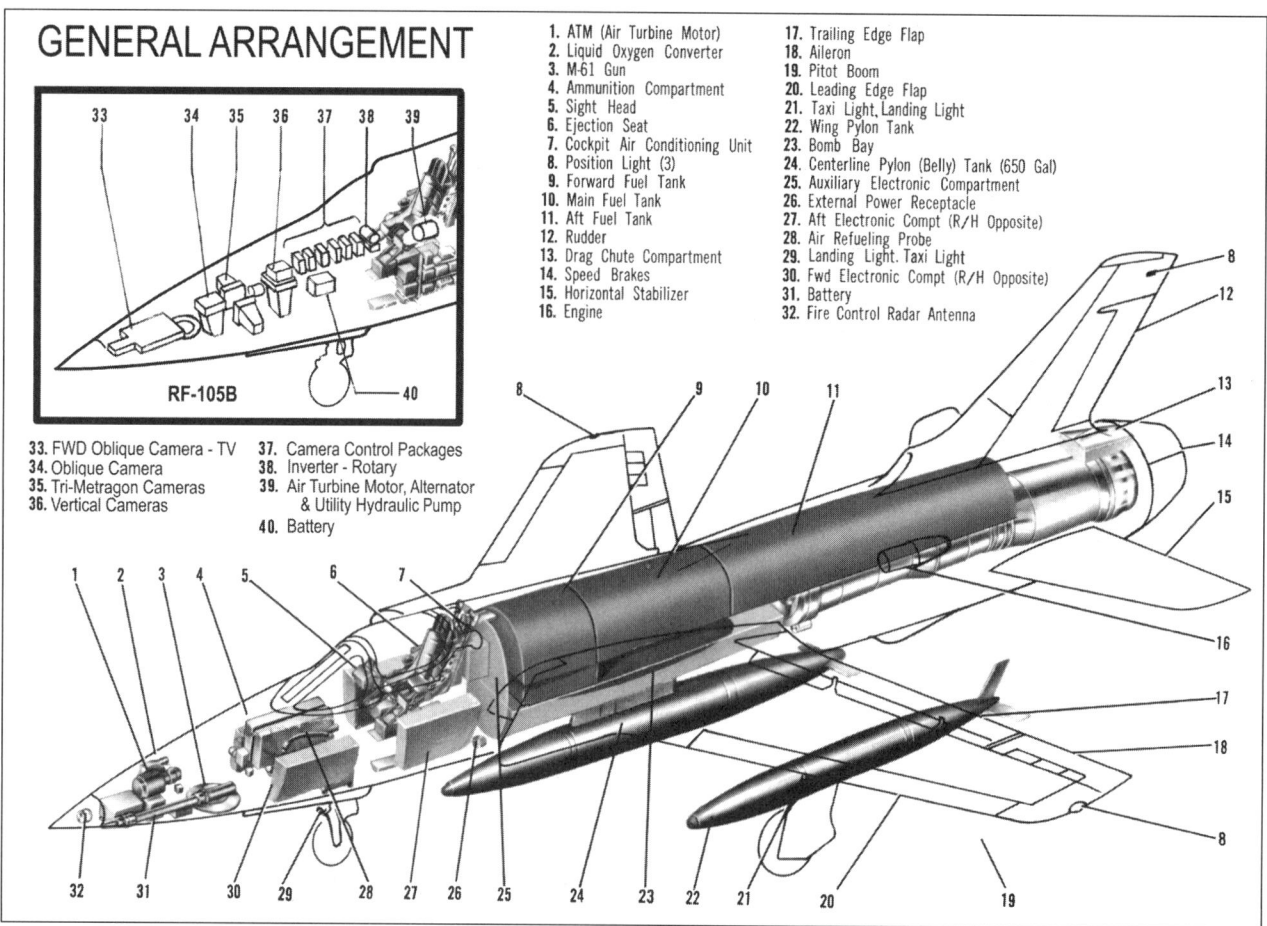

The F-105B flight manual initially included a drawing of the RF-105B camera configuration. By the time the JF-105Bs finally flew, the reconnaissance mission had been assigned to the RF-101. Note the large saddle fuel tank over the J75 engine, a feature of all F-105s. This led to a series of losses when the engine had uncontained turbine failures, and hot blades penetrated the fuel tanks. *(U.S. Air Force)*

weather version was stopped on 27 November, and a week later Republic was told to cease all RF-105 work pending a determination of the project's future.[49]

On 14 January 1957, the Air Force announced the final cancellation of the RF-105 program, with the reconnaissance mission being given to the McDonnell RF-101 Voodoo instead. The three aircraft in production were completed as JF-105B[50] test aircraft without cameras or armament. Without the cameras installed in the modified nose, the aircraft had the volume and weight capacity to carry a significant amount of test equipment and instrumentation, and all three were bailed back to Republic for continued flight-test duties. Republic pilot Lindell Hendrix flew the first JF-105B-1-RE (54-0105) on its maiden flight on 18 July 1957 from Farmingdale. During the next four years, the three aircraft were used in flight evaluations of high-speed flutter, external store separations, and a variety of system tests. At least one of the aircraft (54-0105) has survived and is on display at the USAF History and Traditions Museum at Lackland AFB, Texas. The third aircraft was completed as a JF-105B-2-RE and differed from the first two aircraft in using a J75-P-5 production-type engine instead of the J75-P-3 "50-hour" engine installed in the first two aircraft. All three aircraft were completed with later style canopies (i.e., electrically operated, hinged on each side, with no glazed panels aft), and were retrofitted with J75-P-19 engines.

The first JF-105B-1-RE (54-0105) shows the revised nose contours that were planned for the RF-105 fleet. The camera ports are covered over, and the interior volume used for test instrumentation. The capability to carry a great deal of instrumentation made the JF-105Bs ideal platforms for various systems tests during the early F-105 test program. *(Mike Machat Collection via Tony Landis)*

The last JF-105B (54-0108) also shows the reconnaissance nose contours. At various times the three JF-105s were quite colorful, often seen with checkerboard patterns and yellow or red trim. The unusual fairing located on the side of the fuselage below the windscreen is a rear-looking camera used to film stores jettison tests. *(San Diego Aerospace Museum Collection)*

F-105C

During 1956 the Air Training Command issued a requirement for a two-seat version of the F-105B, developed under the Republic designation AP-63-5 (detail specification ES-347-1). It was to serve primarily as a transition and proficiency trainer, but retained full fighter-bomber capabilities, and was assigned the weapons system designation WS-306L. Externally it was the same size as the F-105B but had a canopy that extended further back to cover a second crew member. The canopy was hinged at the rear, hydraulically actuated, and had a small frame between the two crew stations. The engine was to be the same J75-P-5 originally used by block-10 and block-15 F-105Bs. All the normal B-model avionics, including the MA-8 fire control system with nuclear delivery capability, were retained. The second seat reduced the internal fuel capacity to 1,005 gallons, and the predicted range was down to 561 miles (as opposed to the B's 776 miles).[51]

Contract Change Order No. 1 to letter contract AF-33(600)-32216 was issued on 10 May 1956 to include 5 F-105C-1-REs in addition to the previously specified 65 F-105Bs and 17 RF-105Bs. A full-scale mock-up of the forward fuselage and cockpit was inspected on 15–16 November 1956. A second inspection was conducted 28 February through 1 May 1957 to evaluate the cockpit lighting arrangement and the instrumentation configuration. First flight was scheduled for September 1958.[52]

In a rather complicated contracting move, during October 1957 the Air Force definitized the F-105C portion of letter contract AF-33(600)-32216 with a real contract, AF-33(600)-34496. The Air Force then immediately canceled this contract in its entirety and issued letter contract AF-33(600)-34752 that called for 45 F-105Cs. Within a few days the Air Force canceled the F-105C order entirely and notified Republic that there would be no additional F-105Bs pur-

This is the only known photo of the F-105C cockpit mockup. Unlike the later F-105F, the C-model was the same length as the single-seat F-105B. Both crew members were housed under one canopy that was hinged at the rear. To make room for the second crew member, internal fuel capacity was reduced to 1,005 gallons, reducing the predicted range to 561 miles (as opposed to the B's 776 miles). *(Cradle of Aviation Museum via Ken Neubeck)*

chased past the 78 already on order. It then adjusted the letter contract through termination actions and changes to call for the delivery of 20 all-weather F-105Ds and 8 F-105Es. The D-models were scheduled for delivery beginning in April 1959 and ending in January 1960, while the two-seat E-models would be delivered between July 1959 and January 1960.[53]

Thunderbirds

At the end of the 1963 demonstration season, the Air Force decided that the Aerial Demonstration Team would switch from the North American F-100C Super Sabre to the F-105B, becoming the world's first Mach 2 demonstration team. The Thunderchief would be the third (and last) Thunderbird aircraft from Republic Aviation, following the F-84G Thunderjet and F-84F Thunderstreak. The Thunderchief would have the dubious distinction of having the shortest Thunderbird assignment on record.

The squadron received nine F-105Bs (one block-10, seven -15s, and one -20, all equipped with J75-P-19 engines), with the last arriving on 16 April 1964, just 10 days prior to the beginning of the 1964 season. The aircraft had been pulled from service with the 4th TFW at Seymour-Johnson in October 1962 and ferried to the Republic plant for modification and painting. Equipment unessential to the role of a demonstration aircraft, including the M61 cannon and ammunition drum, APN-105 Doppler navigation system, toss-bomb computer and sight amplification system, E-34 fire control radar and antenna, and APS-54 radar warning system were removed and replaced by ballast.

Two smoke oil tanks were installed in the ammunition compartment, stowage provisions for the pilot's suit bag were provided in the right-side aft electronics compartment, a high-pressure oxygen storage bottle was installed in the left-side electronic equipment

Two of the Thunderbird F-105Bs (57-5783 and 57-5787) practice over Nellis AFB in the Spring of 1964. The four aircraft modified to fly in the slot position had stainless steel vertical stabilizers fitted in order to withstand the heat from the leader's J75. Although the aircraft were only used by the Thunderbirds for part of one season, there were three variations to the fuselage paint scheme, the ones shown here being the last. *(U.S. Air Force)*

compartment, a spare drag chute was carried in the gun breech compartment, a pilot debarkation ladder was stowed in the left console, and new VHF-101 communication and VOR-101B navigation systems were installed in the left forward electronics equipment compartment. The pilot debarkation ladder consisted of two lengths of nylon webbing with seven rungs and a ring on the upper end of each webbing for attaching the ladder to the left longeron. The ladder was provided only in case the aircraft landed at a commercial airport that did not have appropriate utility ladders.[54]

The Air Force always indicated that the Thunderbird aircraft retained a "limited" combat capability, but without a fire control system it is very unlikely the aircraft could ever have participated in combat effectively. The flight manual indicates that the aircraft were still fitted with the appropriate arming and release systems for both nuclear weapons and a range of free-fall bombs. Rocket pods and external fuel tanks could also be carried.[55]

A modification to the flap system allowed deployment of 4° of flaps while maneuvering at speeds in excess of 500 knots with the MASTER SHOW switch on, compared to the normal flaps, which could not be operated at speeds above 280 knots. The fuel and vent systems were modified to allow up to 15 seconds of inverted flight. The variable inlet system duct plugs were deactivated and locked in their aft position, although the bleed doors remained operational. The petals around the afterburner were reprogrammed to allow faster afterburner lights when the MASTER SHOW switch was on. This modification allowed the afterburner to light in under 2 seconds, as opposed to the 3.5- to 5-second delay that was normal for the J75 series. When the MASTER SHOW switch was off, most aircraft systems operated in the normal F-105B way.

The BOMB/ROCKET switch on the control stick was replaced with a smoke tank color selector, and the SMOKE RELEASE switch was located on the throttle grip. The Thunderchief carried two 50-gallon smoke oil containers located in the former 20-mm ammunition bay instead of the single tank carried by previous (and subsequent) Thunderbird aircraft. The aft tank always contained clean oil, which resulted in white smoke. The forward tank could contain a variety of liquids, including dyed carbon tetrachloride, to produce either red or blue smoke. The two tanks fed a single distribution line. When the pilot released smoke, liquid under pressure impinged on the engine exhaust gases, resulting in a smoke trail. Initial smoke delivery took 5 seconds from the pilot depressing the switch, although subsequent smoke deliveries took less than 1 second. A change in color took approximately 5 seconds to purge the line and reprime the system, and both tanks had to be empty prior to landing. Smoke could only be delivered during level or positive-g flight.[56]

A show system control panel (shaped like a Thunderbird, no less!) on the upper left instrument panel contained three switches and eight status lights. The MASTER SHOW switch provided power to the special systems (flaps, afterburner, and smoke) and activated the main fuel tank inverted flight booster pump. A 4° FLAP switch controlled the extension of the flaps in the show configuration.[57]

The KNIFE EDGE switch, which could only be used after the MASTER SHOW switch was activated, deenergized the normal automatic rudder stop and allowed the rudder to travel through its full 32° at all airspeeds. This switch was functional only on the solo aircraft that had a "knife edge" modification to their vertical stabilizers and rudders. These modified aircraft were identifiable by two evenly spaced vertical rows of rivets on the vertical stabilizer tip. The other aircraft, and normal production F-105Bs, had smooth fin caps. Four aircraft also had their vertical stabilizers replaced by units made of stainless steel, allowing them to deal with the exhaust of the leader while in the slot position.[58]

The flight manual listed several restrictions on the aircraft. Knife-edge maneuvers were not to be performed for at least 2 minutes after flying in the slot position to allow the vertical stabilizer to cool. Knife edge flight could only be performed when less than 3,300

The F-105 wore the most elaborate belly paint scheme so far. One of the F-105Bs is shown here on a delivery flight on 25 January 1964 with Major David Pilton at the controls. The Thunderbird looked particularly aggressive, mainly because of the forward-swept air intakes. Unlike previous Thunderbird aircraft, the F-105 was fitted with a smoke system that could dispense either white or colored (normally blue or red) smoke. An unfortunate accident during the seventh show of the 1964 season spelled the end of the F-105 as a demonstration aircraft. *(Cradle of Aviation Museum via Ken Neubeck)*

pounds of fuel remained, and never above 500 knots or 0.9 Mach. The minimum distance between the slot aircraft (top of the canopy bow) and the lead (trailing edge of the bottom speed brake) was limited to 5 feet vertical and 15 feet longitudinally at MIL power, and 14 feet vertical and 15 feet longitudinally in afterburner. The total time the slot aircraft was subjected to the wake immersion from the lead was limited and had to be tracked carefully during each show and practice.[59]

By the beginning of the 1964 show season, the F-105B stood ready to continue in the proud tradition of its predecessors. Six shows were performed beginning with a demonstration in Norfolk, Virginia, on 26 April 1964. During the following two weeks, the team flew performances at NAS Pensacola, Florida; Fayetteville, North Carolina; Patrick AFB, Florida; Shaw AFB, South Carolina; and McChord AFB, Washington. An accident during the 9 May 1964 show at Hamilton AFB, California, killed Captain Eugene Devlin (left wing) when a major fuselage splice plate failed and the aircraft (57-5801) broke apart just in front of the bomb bay. A grounding of all F-105 aircraft resulted. The accident was eventually traced to a primary structural splice plate at the top centerline of the fuselage that had failed due to fatigue. The structure had snapped, then the fuselage broke just forward of the leading edge of the wing. The grounding was lifted in mid-July after a redesigned splice plate had been installed as part of Project Back Bone on all B-models and early D-models that had used the same construction technique.

Pending the outcome of the investigation, the Air Force decided to reequip the squadron with North American F-100D Super Sabres. The plan was to reequip the Thunderbirds with more extensively modified F-105Bs for the 1965 season, but for a variety of reasons this did not happen. The eight remaining aircraft had their systems restored to an operational configuration and were assigned to the 108th TFW, although this did not occur quickly and as late as October 1966 the aircraft were still at the Sacramento depot in the Thunderbird configuration. This was largely the result of rework necessary to keep the combat-capable F-105D/Fs operational. The aircraft that had been modified with stainless steel tails retained them for the remainder of their service careers.[60]

Notes

21. *The F-105: A Chronology (1951–1973)*, History Office, Mobile Air Materiel Area, Brookley AFB, Alabama, September 1963, pp 30–52.
22. *Ibid.*, pp 30–52.
23. *Ibid.*, pp 42–46.
24. *Ibid.*, pp 53–67.
25. *Ibid.*, pp 65–71.
26. History of the Air Force Plant Representative, Republic Aviation Corporation, Farmingdale, Long Island, New York, 1 July 1957 through 31 December 1957, p 77.
27. This was originally intended to be the F-105B, but in reality most of the requirements were deferred to the F-105D.
28. The only one that would find use on the F-105 was the AGM-12 Bullpup, which was designated GAM-83 at the time. The GAM-83A had a conventional warhead; the GAM-83B had a nuclear payload. The requirement for the F-105 to carry the GAM-83B was cancelled on 9 March 1962. The identity of the second type could not be ascertained from available documentation.
29. Knaack, Marcell Size, *Post-World War II Fighters, 1945–1973*, Office of Air Force History, United States Air Force, 1986, pp 192–196.
30. See the separate discussion of the RF-105 for more information.
31. *The F-105: A Chronology (1951–1973)*, History Office, Mobile Air Materiel Area, Brookley AFB, Alabama, September 1963, pp 63–65.
32. Knaack, Marcell Size, *Post-World War II Fighters, 1945–1973*, Office of Air Force History, United States Air Force, 1986, p 193.
33. This item included a note that indicated it could also be carried internally, although no further evidence of this capability has been uncovered.

34. *The F-105: A Chronology (1951–1973)*, History Office, Mobile Air Materiel Area, Brookley AFB, Alabama, September 1963, p 74.
35. *Ibid.*, p 87.
36. Knaack, Marcell Size, *Post-World War II Fighters, 1945–1973*, Office of Air Force History, United States Air Force, 1986, pp 193–194.
37. *Ibid.*, p 193.
38. *The F-105: A Chronology (1951–1973)*, History Office, Mobile Air Materiel Area, Brookley AFB, Alabama, September 1963, p 95.
39. The buddy refueling system was deactivated by TCTO 1F-105-728.
40. Report No. FHB-19-1, Interim Flight Manual, USAF Series F/JF-105B Aircraft, AF Serial No. 54-0100 thru 54-0112, 15 October 1959, P 1–8.
41. *The F-105: A Chronology (1951–1973)*, History Office, Mobile Air Materiel Area, Brookley AFB, Alabama, September 1963, p 125.
42. Knaack, Marcell Size, *Post-World War II Fighters, 1945–1973*, Office of Air Force History, United States Air Force, 1986, p 194.
43. *Ibid.*, p 195.
44. *Ibid.*, p 195.
45. *The F-105: A Chronology (1951–1973)*, History Office, Mobile Air Materiel Area, Brookley AFB, Alabama, September 1963, p 162.
46. Archer, Robert D., *The Republic F-105 Thunderchief*, American Aircraft Series #1, Aero Publishers, Inc., Fallbrook, CA, 1969.
47. Davis, Larry, and Menard, David, *Republic F-105 Thunderchief*, Volume 18 in the WarBird Tech Series, Specialty Press, North Branch, MN.
48. *The F-105: A Chronology (1951–1973)*, History Office, Mobile Air Materiel Area, Brookley AFB, Alabama, September 1963, pp 23–24.
49. *Ibid.*, pp 41–49.
50. Some sources list them as TF-105B, although there is no official record to indicate this designation was used other than a few references in *Post-World War II Fighters, 1945–1973*. The JF-105B designation became official on 14 March 1957.
51. Knaack, Marcell Size, *Post-World War II Fighters, 1945–1973*, Office of Air Force History, United States Air Force, 1986, p 192.
52. History of the Air Force Plant Representative, Republic Aviation Corporation, Farmingdale, Long Island, New York, 1 July 1956 through 31 December 1956, p 59; and 1 January 1957 through 30 June 1957, p 61.
53. History of the Air Force Plant Representative, Republic Aviation Corporation, Farmingdale, Long Island, New York, 1 July 1957 through 31 December 1957, p 12.
54. T.O. 1F-105B(I)-1, USAF Series F-105B Thunderbird Partial Flight Manual, 1 June 1964, changed 15 May 1965.
55. *Ibid.*
56. *Ibid.*
57. *Ibid.*
58. *Ibid.*
59. *Ibid.*
60. Knaack, Marcell Size, *Post-World War II Fighters, 1945–1973*, Office of Air Force History, United States Air Force, 1986, p 195.

Real Thunderchiefs— The F-105D/F

Planning for the F-105D began in the summer of 1957 based on an Air Force request to convert the Thunderchief into an all-weather attack aircraft. The F-105B was still two years away from full squadron service, but it was becoming increasingly apparent that the F-105 needed additional navigation and attack systems to operate in adverse weather. On 6 November 1957, Air Force Headquarters authorized the release of funds to install the Thunderstick fire control system into the F-105. The requirements would be formalized on 22 November with the release of GOR-49-1, and the F-105D and F-105E designations became official on 29 November 1957.[61]

It was also evident that a new engine would be required to maintain an equivalent level of performance, since the new electronics systems would add several thousand pounds. Luckily, Pratt & Whitney had just adapted water injection to the J75-P-19 engine that powered late-model F-105Bs, adding 2,000 lbf for short periods of time. Although the new engine installation would still require a partial redesign, it was the same size and close to the same weight, so major work would not be required. Nevertheless, the air intakes had to be modified to provide additional airflow, changes made to the aft fuselage to accommodate the water injection system, and a stronger landing gear with larger brakes had to be added to handle the increased gross weight of the new design. An arresting hook was added under the rear fuselage, integrated into the ventral fin. This was intended to assist damaged aircraft recovering to land bases and was never intended to allow the F-105 to operate from aircraft carriers.

The F-105D required additional space for the new radar system, beyond that provided by a 15-inch longer nose. Since further extending the nose was undesirable from an aerodynamic viewpoint, the only volume realistically available was the 20-mm link and shell storage area. In the F-105B there had been a dual feed system, with two feed chutes guiding the shells into the breach. The links and empty shell casings were retained after firing and stored in separate compartments. The rationale for this arrangement was to minimize the effects on the center of gravity of firing 500 pounds of ammunition. The feed system was subsequently redesigned, with a linkless belt moving the shells from their storage drum to

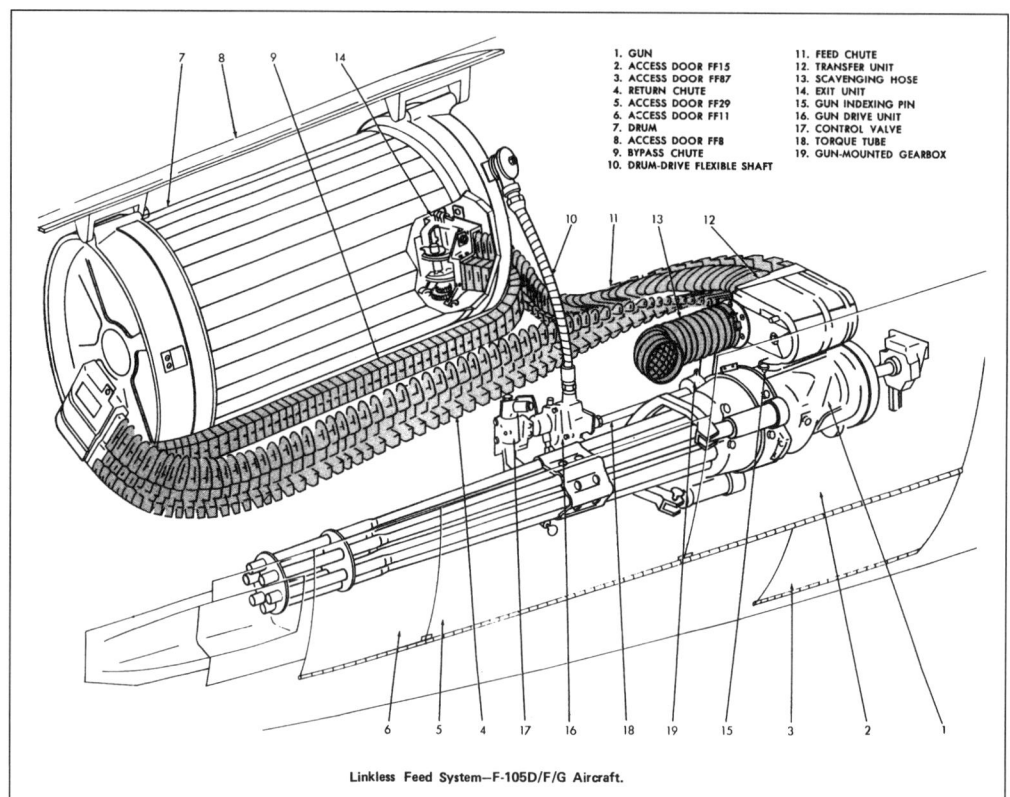

Linkless Feed System—F-105D/F/G Aircraft.

To make room for the new ASG-19 Thunderstick fire control system, the feed mechanism for the 20-mm cannon was completely reworked. Instead of dropping the empty shell casings into a separate container, the casings were now returned to the ammunition drum. This recovered approximately 16 cubic feet of volume. *(U.S. Air Force)*

a conveyor that fed them to the breach. Empty shell casings were extracted from the breech and fed back into the drum for storage. This system recovered about 16 cubic feet.

This volume was then used for the AN/ASG-19 Thunderstick fire-control system, which provided the pilot either a visual or blind delivery option for both nuclear and conventional weapons. The ASG-19 included the North American Aviation Autonetics Division R-14A radar, a considerable improvement over the E-34 ranging radar that had been part of the MA-8 fire-control system in the F-105B. The new equipment had search and ranging functions available for both air-to-air and air-to-ground modes. Interestingly, the radar had originally been designed for the F-107 before that project had been canceled in favor of the F-105B. The aircraft was equipped with a General Electric FC-5 flight control system, which operated in conjunction with the R-14A to provide the F-105D with a true all-weather capability. The radar installation also incorporated a terrain guidance mode, which permitted the pilot to let down through bad weather in unfamiliar territory and hug the ground in order to avoid detection. This was not a terrain-following feature as was later incorporated into the F-111, and it required the pilot to manually fly the aircraft based on cues provided by the system. There were also provisions to fit an AN/APS-92 radar warning receiver and ALE-2 chaff pods.

The earlier F-105s had used conventional circular-display instruments, much the same as had been used by most aircraft for the previous 40 years. The Air Force decided that the F-105D should use a new style of instrumentation called "vertical tape," which presented the information with a series of vertical bars. Studies had shown that this type of instrument was much easier to read when the pilot was busy or under stress, such as in bad weather or combat. The concept behind the installation was to present flight information by means of movable lines related to fixed reference lines. The movable lines presented variable information that the pilot managed and matched to the readings of reference values, which could be preset by the pilot, or in some instances, such as instrument landings, by the onboard avionics. Four new instruments made up the center of the F-105D panel. At the top of the normal "T" was the attitude director indicator (ADI), which displayed roll and pitch attitudes, rate-of-turn and slip, and glide-slope displacement for instrument landings. Directly below the ADI was the horizontal situation indicator (HSI) that showed magnetic heading, bearing, command heading, course information, displacement from course, to/from indication from a desired TACAN fix, and distance to/from the fix. To the left of the ADI was the airspeed Mach indicator (AMI), which presented safe speed warnings, vertical g-loads, true Mach number, and the calibrated airspeed. On the right of the ADI was the altitude vertical velocity indicator (AVVI) displaying vertical velocity (rate-of-climb), pressure altitude, and target altitude.

The cockpit mock-up (station 300 forward) for the F-105D was presented to the Air Force on 10–11 December 1957 and again on 20–21 March 1958. A total of 28 requests for alterations (RFA) were submitted, 15 of which concerned cockpit lighting and 13 for the radar installation.[62] A production go-ahead for the first 28 F-105Ds was given after the March inspection under contract AF-33(600)-34752. By May 1958, Air Force plans called for a 4-year production run of 383 F-105Ds and 89 two-seat F-105Es. The two-seat model was projected to cost about $100,000 more than the single-seater due to extra instrumentation,

While the F-105B had used the conventional round instruments that had been standard for nearly 40 years, the F-105D/F were equipped with a new type of instrument known as vertical tape. This was an attempt to ease the workload on the pilot by making speed and altitude data available quickly, referenced against preset (by the pilot or the flight control computer) values. This photo is of a Thunderstick II-equipped aircraft, although there was very little instrument panel difference from a normal D-model. Note the SST-181X Combat Skyspot transponder control panel on top of the glare shield. *(Mick Roth)*

a second ejection seat, etc. By January 1959, budget restrictions had reduced the quantities to 276 D-models and 83 F-105Es at a combined production rate of 15 per month.[63] On 1 April 1959, the Air Force canceled the procurement of the two-seater and ordered an equivalent number of F-105Ds. It was hoped that this substitution of the less-expensive single-seater would ease the budget problem.[64]

Externally the new aircraft was very similar to the F-105B except for a revised nose to house the new radar. All important dimensions were the same with the exception of the 15-inch longer nose, but many details differed considerably, and the complexity was greatly increased. This is best shown by the fact that it was projected to take 214 workdays to assemble a D-model, while it had only taken 144 workdays to assemble an F-105B. Unfortunately the Air Force could not really afford this increase in labor costs, in addition to the added expense of the new systems. In an attempt to get costs back under control, the Air Force convened a board that included representatives from the three major tactical commands (TAC, USAFE, and PACAF), as well as Air Force Headquarters and Air Materiel Command. The board considered the role of the F-105, its systems, capabilities, and performance, and recommended that the M61 cannon, explosion suppression system for the fuel tanks, APS-54 radar warning receiver, ALE-2 chaff dispenser, ALQ-31 ECM pod provisions, and some minor miscellaneous systems be deleted. It was estimated this would save about $105,000 per aircraft. In the end, operational commanders convinced the board to omit only the APS-92 and some minor equipment.

The three F-105D-1-REs (58-1146/1148) were built intermixed with the last block of B-models to allow them to enter the test program at the earliest possible point. This involved extensive hand assembly since the more complicated and slightly longer D model was not fully compatible with the F-105B assembly line. The first F-105D (58-1146) made

The first F-105D-1-RE (58-1146) poses in front of a Republic hanger. Like the B-models before them, the early F-105Ds were delivered in natural metal finish. This would later be replaced by an overall silver paint, and even later by combat camouflage. The first three D-models were built intermixed with the last of the B-models in an effort to get them into the test program faster. *(Republic via the San Diego Aerospace Museum Collection)*

its maiden flight on 9 June 1959 at Farmingdale with Lin Hendrix at the controls. The only items of note during the test flight were that the vertical tape instruments and new nose-wheel steering worked well.

The aircraft was accepted by the Air Force in July without a complete fire control system but was still considered useful for the first part of the planned test program. On 22 January 1960, the aircraft arrived at Eglin AFB and testing began shortly thereafter. Pilots generally liked the new instrument layout and reported that the handling qualities of the new aircraft were very much like the earlier F-105B. Nevertheless, flight testing of the F-105D proceeded at a slow rate, the various test phases seeing a continuation of the troubles that had plagued the B-model. The new J75-P-19W encountered problems with both the compressor and hot sections, and speed restrictions had to be placed on the aircraft. At one point the problems were so serious that Project Rivet Ram was set up by Air Force Headquarters specifically to monitor progress in finding and correcting the anomalies. The engine problems would continue until late 1968 when new components were finally fitted to all J75s. During the intervening period, failures of the J75 were the leading noncombat cause of F-105 losses.

Just as disturbing were continued problems with the ASG-19 fire control system. A special series of tests, Operation Prove Out, eventually identified modifications that would go a long way toward correcting the reliability and accuracy concerns with the ASG-19. To avoid having a variety of configurations in the growing F-105 fleet, the Air Force decided to incorporate all of the ASG-19 modifications into a single package known as Project Black Box.

On 20 June 1960, the Air Force issued a letter contract to Bendix for the initial manufacture of 85 AN/DPN-61 Homing Set Radars to equip the F-105D with an enemy ground radar

Four F-105Ds show the four-petal speed brake around the afterburner, the same arrangement used on the B-model. The top and bottom petal of the speed brake could only be used in flight, not while landing. This was because the bottom petal could easily scrape the ground, while the top petal interfered with the deployment of the braking parachute. *(Mike Machat Collection via Tony Landis)*

"busting" capability. This was apparently the first modern ECM equipment programmed for the F-105. On 7 December 1960, a team from the F-105 project office concluded that the addition of the DPN-61 was a relatively costly investment with limited dividends. Also, the desired IOC date of December 1961 could not be met. The team based these conclusions on the fact that a lack of all-weather capability would limit its usage, and the technical problems of providing adequate antennas on the aircraft might interfere with the effective use of the radar homing system. Therefore the project office recommended discontinuing efforts to install the DPN-61 in the F-105D aircraft, and the program was terminated on 2 February 1961. Bendix had two sets half-completed when the termination order came. Terminating this program saved the Air Force about $15,000,000, although one of the DPN-61 units would make a short return as part of the equipment on the Wild Weasel IA F-105D (61-0138).[65]

Five aircraft were scheduled to participate in 18 months of Category II testing conducted by the 335th TFS at Eglin AFB. Although the tests were scheduled to start in May 1960, they did not actually begin until 26 December 1960 after the aircraft were modified with the Black Box package. The first modified aircraft arrived at Eglin on 27 October 1960, and the package was cleared for installation in all aircraft during November. However, Republic and other vendors had difficulty producing the required parts for Black Box, and the modification program soon encountered delays that affected production schedules and delayed the flight testing program. All of this led the Air Force to cut short the Category II testing, focusing on the new instrumentation and the fire control systems. It was quickly discovered that when the modified ASG-19 system worked, it worked very well and provided a very credible visual and blind bombing capability. But it was still proving much less reliable than desired when Category II testing concluded on 31 October 1961.[66]

For Category III testing, the 335th moved back to its home base at Seymour-Johnson AFB, consolidating most F-105 activities and support at one location. When the squadron returned home, it took 86 cargo flights to carry 1,007,969 pounds of material and 315 passengers. Despite the reliability problems still being encountered by some systems, Category III testing went well, with few of the support problems encountered with the F-105B. Even the ASG-19 seemed to operate more reliably at Seymour-Johnson, a condition in retrospect that is at least partially attributable to the decrease in humidity at the base compared to Eglin. Many of the reliability problems would return when the aircraft deployed to Southeast Asia. Testing was also still being conducted at Eglin, with three F-105Bs remaining to finish Project Doppler, a test of the APN-105 Doppler navigation unit. By 29 August 1960, these three aircraft had accumulated 185 flight hours and an additional 35 ground operating hours with a 93.7 percent mission success rate. On average the system operated 55.14 hours between failures and had a navigation error of 1.54 percent.[67]

The first operational F-105Ds were delivered to the 4520th Combat Crew Training Wing (CCTW) at Nellis AFB in March 1961. The transition syllabus included almost 40 hours dedicated to the complexities of the new fire control system. The lack of a suitable two-seater again hampered training, but a decision to install the Thunderstick system into six Rockwell T-39s alleviated part of the problem. At this point it was intended to equip 14 TAC wings and two USAFE wings with the F-105D. However, sweeping changes were made by the new Kennedy administration in 1961, and the number fell to 8 wings, the others receiving the F-4 instead.

Worldwide deployment of the F-105D began on 12 May 1961 with delivery of the first aircraft to the 36th TFW, followed shortly thereafter by the 49th TFW. The 36th and 49th TFW were based at Bitburg and Spangdahlem, respectively, as part of the 4th Allied Tactical Air Force (ATAF), United States Air Forces Europe (USAFE). The ferry flights were known as Project High Flight, and the last of 203 F-105Ds were flown from Brookley AFB to Europe on 24 January 1963. The Thunderchief was publicly presented to the Europeans at the Paris Air Show on 3 June 1961.

This F-105D-5-RE (59-1719) shows how the antiglare shield was painted all the way back past the cockpit. This made the aircraft look even sleeker than it already was. There were variations in the color of the aluminum alloys, giving the aircraft a somewhat mottled appearance. *(Mike Machat Collection via Tony Landis)*

An F-105D-5-RE (58-1158) displays the Indian-head motif on the forward fuselage and the Tactical Air Command shield and lightning bolt on the tail. While assigned to squadrons that routinely stood nuclear alert in Europe and Japan, the F-105s generally carried a drop tank under each wing, leaving the weapons bay available to carry a tactical nuclear weapon. *(Peter M. Bowers)*

One example of the F-105s performance was demonstrated soon after F-105D Category II testing began. Lieutenant Colonel Paul Hoza, commanding officer of the 335th TFS, flew a new F-105D nonstop from Eglin to Nellis on 10 July 1961. The 1,520-mile flight was accomplished at altitudes of between 500 and 1,000 feet AGL, and was flown blind all the way. A single aerial refueling was accomplished above western Texas.

On 16 July 1961, the first detachment from Bitburg arrived at Wheelus AB, near Tripoli, Libya, to conduct tactical training. Wheelus, the USAFE center for fighter-bomber and interceptor tactics, included a syllabus on special store delivery. The Wheelus range was so well utilized that at any given time about 10 percent of the available USAFE F-105s were there. Some 50 percent of the total flying time for the squadrons was recorded at Wheelus, or in transit. Even with aircraft out of service while undergoing Look Alike modifications, the two USAFE wings logged 33,000 flight hours during their first 16 months with the type.

The Pacific Air Forces[68] (PACAF) 8th and 18th TFWs were equipped beginning in October 1962, with aircraft delivered as part of Project Flying Fish. By 31 May 1963, 104 F-105Ds had been delivered to Kadena AB, Okinawa, and Japan.

The F-105D was originally intended for the nuclear strike role, with the primary armament being a "special" (nuclear) weapon housed in the internal weapons bay. This weapon was usually an Mk 28 or an Mk 43. However, an Mk 61 could be carried underneath the left or right inboard underwing pylon, and an Mk 57 or an Mk 61 could be carried underneath the centerline pylon. As nuclear war became less and less likely, the nuclear weapon in the

Project Look Alike replaced the natural metal finish with an aluminized silver acrylic lacquer in an attempt to better seal the aircraft against moisture. It also brought all F-105Ds up to the block-25 standard. A continuation of Look Alike would later bring all D-models up to the block-31 configuration. Here a block-20 aircraft (61-0127, foreground) and a block-31 (62-4367, background) take on fuel from a KC-135. *(U.S. Air Force/DVIC)*

internal weapons bay was usually replaced by a 390-gallon internal fuel tank. From block-20 onward, the F-105D had increased emphasis on conventional weapons delivery, with the capability of carrying a wide-ranging combination of external stores underneath the wings and fuselage. Earlier F-105Ds were eventually upgraded to this standard. In June 1961, the F-105D demonstrated its capability to deliver 7 tons of bombs during special tests at Eglin AFB, the heaviest load of bombs ever carried by a single-engined fighter. This feat was repeated in October 1961 at Fort Bragg, North Carolina, during a demonstration for President Kennedy. The F-105's maintenance requirements began to ease in September 1961 when the maximum time between overhauls for the J75 was raised from 300 hours to 600.

However, during December 1961 all F-105s were grounded for inspection after a main forged-steel frame at fuselage station 442.785 failed during laboratory fatigue tests at Wright-Patterson AFB. Further testing confirmed that the frame retained considerable strength after cracking, but any aircraft discovered with cracks was restricted from flight duty. Republic developed a modification to the frame that was quickly incorporated onto the production line, and earlier aircraft were returned to the factory to be brought up to the new configuration.[69]

Although GOR-49-1 had specified the capability to launch the Maxson GAM-83 (AGM-12) Bullpup, this capability was not incorporated in the initial F-105Ds. Provisions for launching the conventional version (mainly wiring) were included beginning on the 70th aircraft, but the actual equipment to do so did not appear until the 115th aircraft. This aircraft also included the capability to carry the nuclear-armed version of Bullpup, and these capabilities were later retrofitted to all F-105Ds.[70]

As the F-105D project was initially authorized, there were four block configurations. The first three aircraft were block-1 and contained minimal systems for flight testing. The next 65 aircraft were block-5 and contained the baseline ASG-19 fire control system. Block-6 would add a blind dive toss capability, provisions for the AGM-12 Bullpup, a rocket-powered ejection seat instead of the catapult unit, and a gallons-per-mile fuel indicator. Block-10 would add capabilities for a 650-gallon centerline tank, a 200-gallon wraparound bomb bay tank that still permitted carrying a single nuclear weapon, and full AGM-12 capabilities. Block-10 also deleted the APS-54 radar warning system and the ALE-2 chaff dispenser provisions, and these systems on earlier aircraft were deactivated.

At the time the F-105 was designed, TAC operated its own probe-and-drogue tankers such as converted KB-50s, whereas flying-boom tankers such as KC-97s and KC-135s were assigned to SAC. The Navy also operated probe-and-drogue tankers such as the KA-3. The F-105 was fitted with a retractable refueling probe, which swung outward from a slot in the forward fuselage just ahead of the cockpit. Unfortunately, the retirement of the KB-50 tankers left TAC dependent on SAC's boom tankers. Although a hose extension could be attached to the end of the flying boom so that it could handle probe-equipped receiving aircraft, it was thought that it would be beneficial for TAC to have aircraft that were fully compatible with SAC booms. Project Stay On III was initiated on 9 January 1961 to devise an efficient way to refuel the Thunderchief, and the last 135 F-105Ds (block-31) introduced a dual in-flight refueling capability. This was done by fitting a flush-mounted retractable door-type receptacle in front of the windshield that could accept a flying-boom. This allowed the aircraft to take on fuel from either the Air Force-standard flying-boom tankers, or from the Navy-standard probe-and-drogue system. This capability was incorporated into all the subsequent F-105F two-seaters and was retrofitted to earlier F-105Ds as part of time compliance technical order (TCTO) 1F-105D-675 (Mod 1128).

On 2 April 1962 a labor strike at the Republic factory again halted F-105 production. Eventually the Pentagon convinced the president to end the strike using the Taft-Hartley powers, but production was not resumed until 18 June. By then two accidents involving F-105Bs at Nellis had raised serious concerns about the aircraft's safety. On 20 June all

This F-105D-10-RE (60-0457) carries two LAU-32 rocket pods under the wing. Each 10-inch diameter pod carried seven 2.75-inch rockets, instead of the 19 carried by the 16-inch diameter LAU-3 pods that were used in Southeast Asia. *(San Diego Aerospace Museum Collection)*

F-105s were again grounded, and the early D-models were included in the same Project Look Alike modifications being given to the F-105Bs.

For the F-105D, Project Look Alike would eventually bring all F-105Ds up to block-25 standard (and later, the block-31 standard, beginning on 12 October 1962). A total of 385 TCTOs were incorporated into earlier aircraft, and part of the program provided for the quadrupling of the aircraft's capacity to carry externally mounted 750-pound iron bombs. Initially, one 750-pound bomb could be carried on each of the four wing stations, while after the modification six could be carried on a multiple ejector rack (MER) under the fuselage, four each on MERs on the two inboard wing stations, and a single bomb on each of the outboard stations, for a total of 16 bombs. The modification program required, among other things, the expensive and time-consuming process of removing flush-riveted skin panels to cure the same corrosion problem experienced by the F-105B. Like the B models before them, the F-105Ds were then painted in an aluminized silver acrylic-lacquer, which sealed the seams in the skin to prevent a recurrence of the corrosion problem. The entire process took about 1 month per aircraft, and consumed a total of 5.4 million man-hours of direct labor and $51,000,000.

Project Look Alike created its own problems, however. Republic and the Air Force had been struggling to get the spare parts inventory up to acceptable levels. Look Alike basically wiped out this inventory, using the newly procured parts during the modifications. This left the operational units with precious few spares to service their aircraft during 1963 and 1964. Just when this was being overcome, new problems surfaced with the J75. Engine failures, fuel leaks, and malfunctions of the fuel venting system became epidemic, leading to a variety of new modifications. One of the problems discovered was that the original fuel drain was not long enough to get dumped fuel clear of the fuselage boundary layer. Fuel was sucked against the lower aft fuselage and flowed to the end of the speed brake petals, where it encountered a low-pressure turbulent area and was sucked into the area under the

Although the Thunderchief was designed and procured as a nuclear strike fighter, the Air Force quickly recognized its potential as a conventional fighter-bomber. Early in the test program trials were conducted with a variety of conventional weapons, including the sixteen M117 750-pound bombs seen here on an F-105D-5-RE (58-1173). Only four bombs (instead of six) could be carried on the inboard wing pylons because of interference with the main landing gear. *(San Diego Aerospace Museum Collection)*

The Pratt & Whitney J75 engine was the most powerful turbojet of its era, but it proved troublesome in initial service. In fact, engine failures were the second most common cause for F-105 losses after combat. The engine would mature into a responsive powerplant that provided the F-105 with the ability to exit trouble areas quickly, frequently surpassing Mach 1 at low level. *(U.S. Air Force Historical Research Agency Collection)*

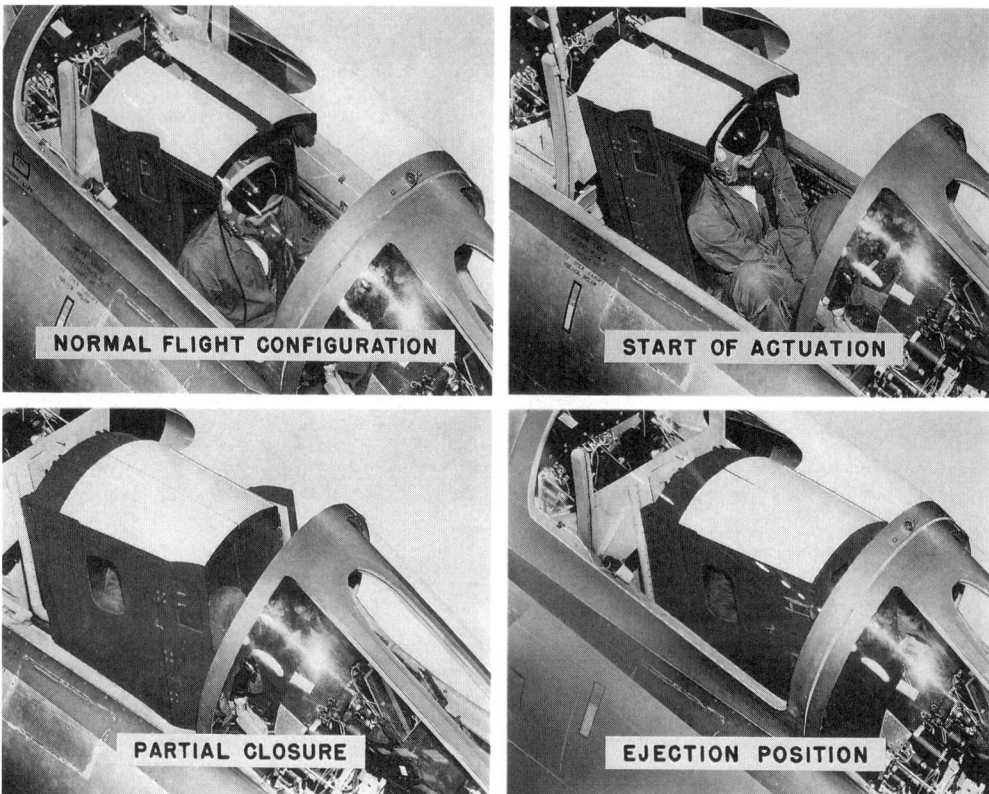

Very little is known about this proposed supersonic ejection system for the F-105 except that it was never installed on any operational aircraft. Although the capsule looks like it would have provided a great deal of protection for the pilot during high-speed ejections, it also looks very heavy and would probably have adversely affected the aircraft's center of gravity and handling. *(Mike Machat Collection via Tony Landis)*

afterburner section of the engine and pooled there. Next time the afterburner was lit, this fuel could be explosively burned off. Another problem was that the turbine blade bracket was coming apart and the blades ejected around the aft section of the engine. This usually resulted in the blades cutting hydraulic lines and puncturing the saddle fuel tanks over the engine, resulting in immediate catastrophic fires. This also led to a shortage of spare engines, hampering F-105 operations from 1964 through the end of 1967.

On 18 March 1963, the Air Force announced that the F-105 had been selected to carry the Shrike antiradiation missile being developed by the Navy. This missile was described as "a supersonic air-to-ground antiradiation missile planned for use against enemy radar installations." Although only of casual interest in 1963, this capability would later see wide use in Southeast Asia.[71]

The F-105 was proving to be a troublesome aircraft in operational service, and one not universally liked by its pilots.

In service, the seven TAC wings using the F-105D experienced a series of accidents and operational problems that led to no less than 10 groundings. As of 31 May 1963, 23 of the 519 F-105Ds had been lost in accidents during a 4-year period—5 during May 1963 alone. In 1964, 38 F-105s were lost to explosions or fires. A more careful examination of the statistics, however, reveals that the F-105D had the second best accident record of any Century Series fighter during its first 50,000 flight hours. Nevertheless, the aircraft came to be known under the rather derogatory nickname "Thud" and had a questionable reputation for being too heavy, having too long a takeoff run, and being a maintenance nightmare.[72]

But when it worked, it worked well—on 15 February 1964, six F-105s set an unofficial record for formation flight when they flew nonstop 4,422 miles from Guam to New Zealand in 7 hours, 45 minutes.

The derogatory comments would largely end in early 1965 when the F-105D began flying combat missions from Korat AB, Thailand. Beginning in 1966 and subsequent years, the F-105 would carry out more strikes against North Vietnam than any other Air Force type. Operating against ever-increasing enemy defenses, the F-105 would also suffer more combat losses than any other U.S. type. The steady loss of F-105s to enemy action, accidents, and normal attrition necessitated urgent repairs and cannibalization of the more badly damaged aircraft, and resulted in a near depletion of USAFE and TAC inventories. TAC's resources for training and support of the combat effort were also reduced.[73]

The sheer magnitude of this can be seen by examining the work performed by the Sacramento Air Materiel Area (SMAMA). The depot at McClellan AFB had taken over from the Mobile Air Materiel Area (MOAMA) at the beginning of 1967. During 1967 the Sacramento depot had 29 major modification programs in-work for the F-105, an increase of 13 from the previous year. These modifications would consume more than 553,000 man-hours of direct labor. In addition, Sacramento performed IRAN (inspect and repair as necessary) work on the F-105. Each of the 126 IRANs performed in 1967 took 7,700 man-hours, for a

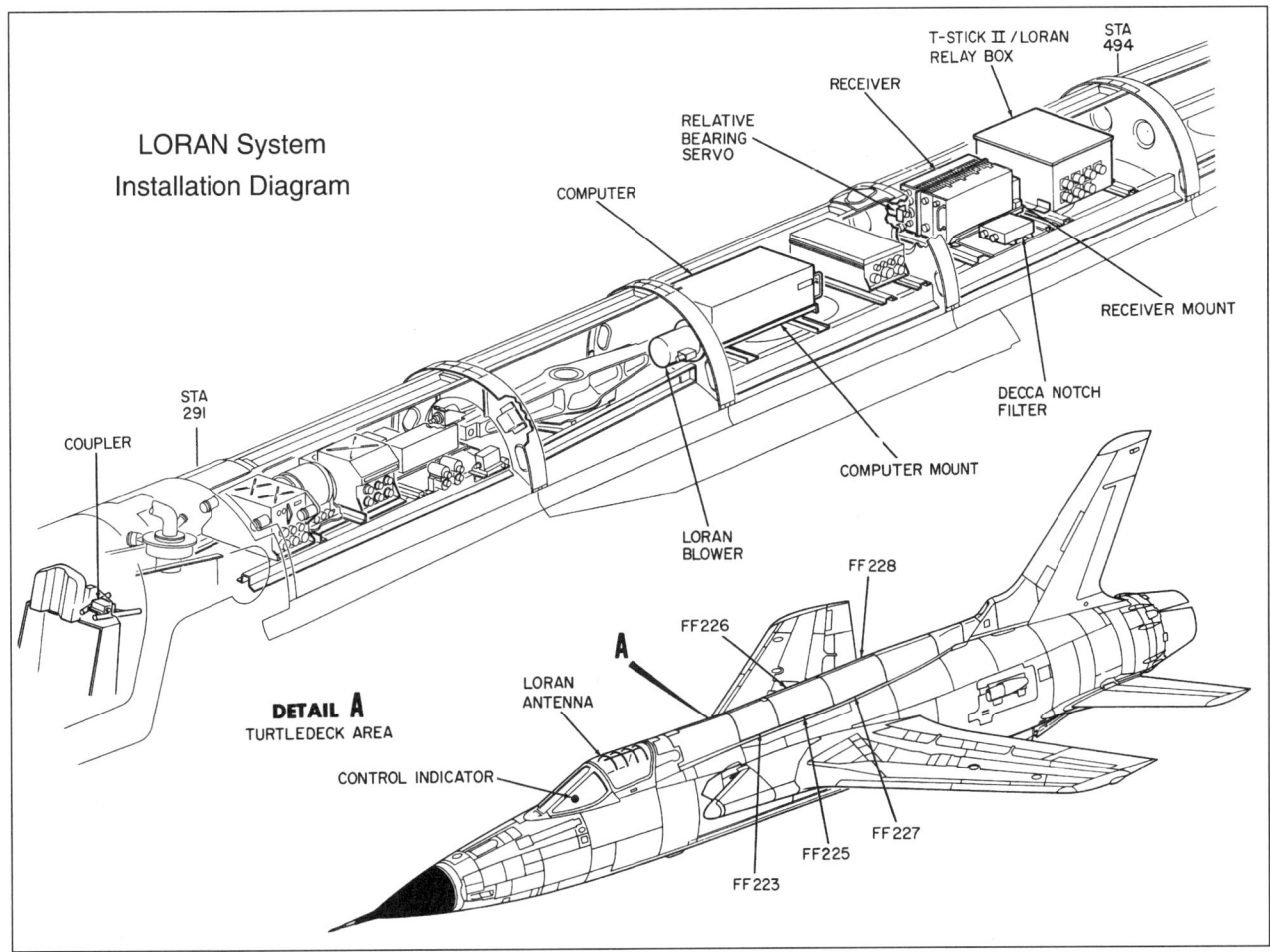

There was simply no space left in the fuselage of the F-105D to house the Thunderstick II's electronics, so the Air Force decided to make additional volume available in the form of a "turtledeck" fairing on top of the fuselage. Similar fairings were also used on the Navy's A-4 Skyhawk and RA-5C Vigilante. Components of the Thunderstick II fire control system, primarily the ARN-92 LORAN, were housed under this fairing. (U.S. Air Force)

total of 970,000 man-hours. By 1967 the first F-105D had accumulated 2,000 flight hours, half of the 4,000-hour design life. And the hours being flown were harder than expected, necessitating major structural repairs on some aircraft.

A total of 610 F-105Ds were built, with 17 delivered in FY60, 149 in FY61, 171 in FY62, 198 in FY63, and the final 75 in FY64. Interestingly, the F-105D is listed as being over 50 percent less expensive than the preceding F-105B, even though it was a much more capable aircraft. This is probably a result of having amortized the majority of development costs with the F-105B production. Each F-105D cost an average of $2,140,000, including $1,472,145 for the airframe, $244,412 for the engine, $19,346 for the electronics, $167,621 for armament, and $19,346 for ordnance. The average cost per flight hour is listed as $1,020, with an average maintenance cost per flight hour of $809.[74]

Thunderstick II

The initial use of the F-105D in Southeast Asia showed that although the ASG-19 was an advanced fire control system, it was not as accurate as had been expected when it was developed. To improve the attack and navigation capabilities of the Thunderchief, 30 F-105Ds were modified to carry the Thunderstick-II bombing system in a large "turtledeck" fairing extending from the rear of the canopy to the base of the vertical stabilizer. Although the F-105D had been intended to be an all-weather fighter-bomber, operational experience in Vietnam showed that its capabilities were marginal at night or in bad weather. Conceived in 1966, the T-Stick-II system was intended to give the F-105D a true blind-bombing capability, as well as to improve its visual bombing from low altitudes. A new ITT AN/ARN-92 LORAN receiver was added in the turtledeck fairing that added 22

The turtledeck fairing did not improve the F-105D's aesthetics. By all reports the Thunderstick II system was a large improvement over the normal ASG-19 in terms of accuracy and reliability. Unfortunately, it came too late to see operational service in Southeast Asia, and only 30 aircraft were modified, including this F-105D-20-RE (61-0161). All 30 aircraft were initially assigned to the 23rd TFW at McConnell AFB, but by 1972 all had been transferred to the 457th TFS, an Air Force Reserve unit at Carswell AFB. *(Mick Roth Collection)*

The Sacramento depot at McClellan AFB completed four T-Stick II aircraft by 31 May 1970, and six more would be completed by the end of June. Surprisingly, the remaining 20 aircraft were modified by Air America at Tainan, Taiwan, beginning in July 1970. All 30 aircraft were finally completed in July 1971, 5 years after they had been selected for the program. *(A. Staruszkiewicz via the Terry Panopalis Collection)*

cubic feet of volume, made no significant change in drag, and actually improved directional stability somewhat. The LORAN antenna was taped to the inside of the canopy, just over the pilot's head. The R-14A radar system had its original electronics replaced with solid-state devices and was redesignated R-14K. A Singer/General Precision gyrocompass attitude vertical reference system worked in conjunction with the AN/APN-131 Doppler navigation unit. Flight tests were begun at Eglin during September 1969 and were completed on 7 November. The Sacramento depot completed four T-Stick II aircraft by 31 May 1970, and six more would be completed by the end of June. Surprisingly, the remaining 20 aircraft were modified by Air America at Tainan, Taiwan, beginning in July 1970. All 30 aircraft were finally completed in July 1971, 5 years after they had been selected for the program.[75]

The aircraft demonstrated that when all its systems were working correctly it could achieve CEPs of less than 50 feet from 15,000-foot altitudes. Perhaps even more impressive, the terrain avoidance feature of the radar was accurate to within 30 feet at altitudes less than 300 feet. However, the limited number of aircraft, and even more limited number of spare parts and trained technicians, kept the aircraft from exploiting its full potential. All 30 aircraft were initially assigned to the 23rd TFW at McConnell AFB, but by 1972 all had been transferred to the 457th TFS, an Air Force Reserve unit at Carswell AFB. Ten years later the last one was ferried to Davis-Monthan AFB for storage and eventual scrapping.

RF-105D

As early as 3 May 1959, there was mention of a reconnaissance version of the all-weather Thunderchief. Amendment No. 2 to GOR-49, published on 7 December 1960, called for an advanced reconnaissance aircraft tentatively designated RF-105D. The concept centered

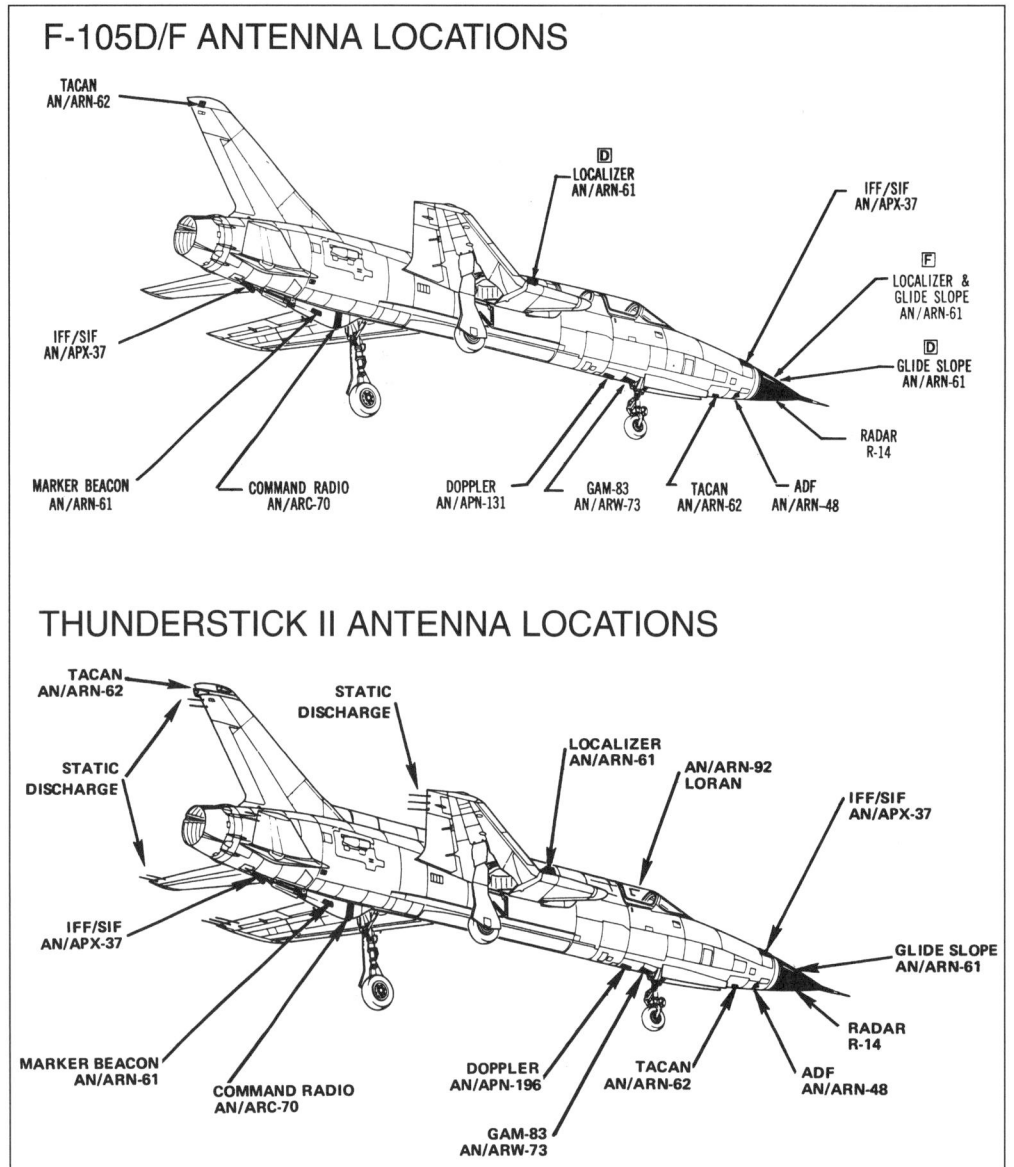

The Thunderstick II aircraft added a LORAN antenna, but were otherwise similar to the normal D/F-models. Interestingly, the aircraft also added three static discharge strips on the vertical stabilizer, ailerons, and stabilators. This drawing does not show the location of the various ECM antennas that all D/F-models received during the war in Southeast Asia. *(U.S. Air Force)*

on an aircraft equipped with a side-looking radar in a centerline pod, infrared sensors, a variety of camera installations, and an extensive electronic intelligence (ELINT) gathering system. The requirements included in-flight processing of films and the ejection of film cassettes directly to field commanders.

But there was a breakdown of communications within the Department of Defense—although everybody seemed to know that an RF-105D was needed, nobody issued a contract for the development of the aircraft or its systems. The confusion regarding the RF-105D project caused the Air Force to request supplemental funding to keep one RB-66 and one RF-101 squadron in service longer than anticipated.[76] On 2 June 1961, Republic finally received a letter contract for the development of the RF-105D. The Air Force indicated that the desired number of RF-105Ds should be sufficient to equip four squadrons by

30 June 1965. Although it is fairly certain that some amount of work was performed on systems development, it is unclear if there was any actual airframe development performed. The project was officially canceled on 29 December 1961, and the SOR 49-2 requirements, along with systems work already performed, were transferred to what eventually became the RF-4C[77] (under SOR 196).[78]

F-105E

When the two-seat F-105C was canceled in October 1957, it was primarily because the emphasis had shifted to the newer single-seat D-model, instead of the B-model the two-seater had been based on. Since the requirement that had led to the C-model being developed was still valid, Republic began developing a two-seat version of the new F-105D under the company designation AP-63-33. The aircraft that emerged was the same size as the F-105D (i.e., it was not stretched like the F model would be) and had a rear-hinged, one-piece, bubble canopy covering the two crew members. Power would be provided by the same P&W J75-P-19W engine used by the F-105D. The aircraft retained all the normal D-model avionics, including the nuclear-capable Thunderstick fire control system and R-14A radar. Additionally, it was envisioned the aircraft would carry the APS-54 radar warning receiver and two external ALQ-31(V) ECM pods. Internal fuel capacity was again reduced to 1,005 gallons, and this coupled with the slightly greater drag of the new canopy served to reduce the combat range of the aircraft somewhat from that of the single-seater. First flight was to be in November 1959.

In June 1958 Republic proposed modifying one of the JF-105Bs to a superficial F-105E configuration to test the aerodynamics of the new forward fuselage and canopy, including spin testing. It appears that no action was taken on this proposal.[79]

An artist's concept of the F-105E. The two-seater was the same length as the F-105D, sacrificing some internal fuel to make room for the second seat. Unlike the earlier proposed C-model, the F-105E used a single-piece transparency over the pilots. Significant parts for the first ten F-105Es were under construction when the program was canceled. *(Cradle of Aviation Museum via Ken Neubeck)*

Republic brought the F-105C forward fuselage mockup out of storage and modified the cockpit to the E-model configuration. Here the forward ejection seat is sitting on a board outside the cockpit so that the new chaff dispenser installation may be inspected during a review on 6 February 1958. The mockup was modified to incorporate the new vertical tape instruments and other ASG-19 controls and displays. *(Cradle of Aviation Museum via Ken Neubeck)*

The F-105E mockup sits ready for the final cockpit lighting inspection. Interestingly, Republic did not modify the forward fuselage of the mockup to resemble the ASG-19 equipped aircraft, and the early B/C-model nose remained until the mockup was disposed of. In this photo the original F-105C canopy is also installed, complete with a two-piece transparency. Production F-105Es would have eliminated the center support between the pilots. *(Cradle of Aviation Museum via Ken Neubeck)*

Contract AF-33(600)-36687 dated 7 November 1958 placed an order for 26 F-105E-5-REs (59-1817/1842). The forward fuselage mock-up that had been constructed for the F-105C was brought out of storage and the cockpits were converted to show the new instrument arrangement, using the same vertical tape instruments as in the F-105D. Interestingly, the external shape of the nose was not changed to reflect the larger-diameter fuselage necessary to hold the R-14A radar. This was largely because the nose was to be identical to the standard F-105D, and Republic elected to save money by not changing the mock-up. A development engineering inspection was held on 3–4 May 1958 and resulted in 27 requests for alterations. Eighteen of these had already been corrected by the time the formal mock-up inspection was held on 26–27 May 1958. Less than a year later, on 1 April 1959, the Air Force canceled the F-105E as a cost-saving measure. It was felt that by not having to divide the production effort on two models the manufacturer could make up time and achieve some financial savings on a project significantly over budget and behind schedule.[80]

By the time the contract was cancelled, Republic was already fairly far along manufacturing major assemblies for the first eight F-105Es. Since, with the exception of the forward section (which contained the cockpit, stations 196–349) these assemblies were identical between the D-model and the E-model, it was decided they should be used to construct single-seaters:[81]

Nose Section (stations 0–195):	Rear Fuselage:
E2 convert to D10	E2 convert to D11
E3 convert to D8	E3 convert to D6
E4 convert to D9	E4 convert to D7
E5 convert to D14	E5 convert to D8
E6 convert to D19	E6 convert to D19
E7 convert to D25	E7 convert to D25
E8 convert to D26	E8 convert to D26
Aft Section (stations 526–633):	**Wings:**
E1 convert to D28	E1 convert to D4
E2 convert to D6	E2 convert to D14
E3 convert to D7	E3 convert to D15
E4 convert to D8	E4 convert to D16
E5 convert to D9	E5 convert to D17
E6 convert to D10	E6 convert to D18
E7 convert to D11	E7 convert to D19
E8 convert to D12	E8 convert to D20

The entire forward fuselage (station 525 forward) from the first F-105E (E1) was used as a fatigue test airframe. The only two center sections (stations 350–525) that were under construction were used in D17 (from E6) and D14 (from E7). In addition, subassemblies were in work for the next 10 two-seaters, and these were folded into the D-model production line. The first 10 serial numbers that had been assigned to F-105Es were used by F-105D-6-REs.

When the F-105E was canceled, the Air Force was faced with finding a method of training the ever-increasing number of F-105 pilots that were required. Initially they intended to use the F-100F for combat proficiency evaluation and transition training, but this presented several problems. First, there were a very limited number of two-seat F-100Fs available, and they were largely needed to train F-100 pilots. Secondly, the F-100 had vastly different flight characteristics from the F-105D, so the training was of limited value. Having rejected this approach, TAC used six modified Rockwell T-39 Sabreliners to train pilots in the use of the R-14A search radar pending a final solution.[82]

Finally, a Two-Seater: The F-105F

As the F-105 got heavier and heavier, the need for a trainer grew steadily—the performance margins were going away, and it was becoming too risky to allow new pilots to solo on their first flight. In May 1962, Secretary of Defense Robert McNamara gave his go ahead to develop a two-seat derivative of the F-105D. In June the Air Force announced its intention to procure 36 two-seat F-105Fs in FY62 and an additional 107 aircraft in FY63. It was expected that the FY62 aircraft would replace a like-number of F-105Ds on the assembly line, but that the FY63 aircraft would be in addition to the F-105s already scheduled for production. Later in 1962, the last 143 single-seat F-105D-31-REs were canceled and replaced by 107 F-105F-1-REs. Regardless, it still cost an additional $8 million to complete these aircraft as two-seaters.[83]

Unlike the previous two-seaters (F-105C and E), Republic decided to stretch the fuselage of the F-model, and a 31-inch plug was inserted ahead of the wing to accommodate the second seat. As a result of the structural change, the F-105D's electronic equipment compartment was eliminated. However, two electronic compartments, one on each side of the aircraft below the aft end of the extended crew compartment were provided and yielded roughly the same usable volume. Then, because the center of gravity had shifted forward, 5 inches were added to the top of the vertical stabilizer and its chord was increased, resulting in 15 percent more area. The center and aft fuselage were reinforced to handle the increased aerodynamic loads from the larger vertical stabilizer. As a result of a 3,000-pound increase in weight, the landing gear was strengthened and modified with a 350-knot retraction capability. The air direction detector (angle of attack vane) was relocated to the right side of the fuselage to preclude any possibility of damage during air-refueling operations.[84] The forward fuel cell was redesigned due to the fuselage changes, but the total fuel capacity was unchanged. The second cockpit was equipped for a pilot, with full flight instrumentation, but only limited armament controls. The canopies incorporated a combined actuator/remover unit for each crew station, in lieu of separate canopy actuator and canopy remover units due to space limitations. Each canopy opened 75° instead of the 42° of the single-seat aircraft, affording an equivalent unobstructed area for normal ingress and egress. The crew survival kits were simplified, and a 750 VA static inverter was added to provide additional AC electrical power. Republic described the function of the two-seater as:

> The two-place F-105F airplane is still considered to be and be employed, in its primary function, as a 'one-man' combat weapon to execute the same missions, from the FWD [forward] crew station, as programmed for the single-place F-105D airplane.... Should the aircraft be sent into combat with an 'extra' man occupying the AFT crew station, execution of the mission will not be contingent upon the 'extra' man performing any function except to actuate one control which constitutes 'consent-to-unlock-and-arm' a nuclear weapon preparatory to release.[85]

This is just as well since the visibility from the second cockpit was too limited to make it a practical conversion trainer, but it could nevertheless serve to introduce new pilots to the Thunderchief and provide a proficiency trainer. It also proved very useful for radar/navigation/bombing training. However, in the end, the F-105F's primary contribution would come from a totally unanticipated use.[86]

A single-seat F-105D-5-RE (58-1155) was ballasted to simulate the center of gravity of the two-seat F-model and evaluated the modified control system destined for the new aircraft. Wind tunnel tests were conducted on a 1/22nd scale model of the taller vertical stabilizer, and as a result the aircraft was cleared for flights up to maximum speed at all altitudes. During June–October 1962, a rocket sled test program was conducted at Holloman AFB, New Mexico, to qualify the modified escape system.

Republic constructed a full-size metal mock-up of the crew compartment and nose section for use as an engineering and manufacturing tool. An F-105F cockpit preview—an unofficial mock-up inspection—was held on 5 June 1962. All of the requests for alteration were incorporated before the formal mock-up inspection was held on 2–5 January 1963. The mock-up was subsequently modified to include functional cockpit lighting, and a cockpit lighting inspection was held 19–20 March 1963.[87]

The first F-105F (62-4412) came off the assembly line on 23 May 1963 and flew on 11 June 1963, some 10 days ahead of schedule. Carlton B. Ardery, Jr., pushed the aircraft to Mach 1.15 on this flight, announcing that it handled pretty much like a D-model. Since the F-105F was essentially a derivative, it did not require an extensive test program. Category I testing was begun in mid-1963 and was completed in July 1964. Category II testing was completed in August 1964. On 7 December 1963, the first F-model was delivered to the 4520th CCTW at Nellis, and on 23 December the first operational aircraft was delivered to the 4th TFW at Seymour-Johnson. During a tour to demonstrate the new aircraft to a series of high-level Air Force commanders, an F-105F completed 61 flights in 22 days, logging 72

All F-105Fs were completed with rocket-powered ejection seats, unlike all B-models and the majority of D-models that were delivered with catapult seats. Most D-models were later retrofitted with the rocket-powered seats that offered much better low-speed/low-altitude capabilities. In preparation for the upcoming two-seat F-model, the Air Force tested the ejection systems on a rocket sled. *(Mike Machat Collection via Tony Landis)*

At long last, a two-seat Thunderchief. Unlike the stillborn C- and E-models, the F-105F was 31 inches longer than the single-seater, allowing all internal fuel to be retained. But the cockpit configuration did not allow much visibility from the rear cockpit, somewhat limiting the aircraft's value as a trainer. It did not really matter since the primary fame for the two-seater would come from an entirely unexpected use. *(San Diego Aerospace Museum Collection)*

Many of the early F-105Fs were delivered to the 4520th Combat Crew Training Wing at Nellis AFB. All the F-105Fs were delivered in the Look Alike silver paint scheme. The F-model used a larger vertical stabilizer to compensate for the longer fuselage but was overall very similar to the D-model. *(San Diego Aerospace Museum Collection)*

The second F-105F-1-RE (62-4413) was retained at Eglin AFB for a variety of tests. This aircraft spent most of its career in an overall light gray paint scheme. Note that the antiglare shield does not extend rearward from the canopies. *(via Paul Minert via the Mick Roth Collection)*

hours, 35 minutes of low-level time, including a supersonic dash and a simulated nuclear weapon delivery.

By December 1964, the last Thunderchief was slowly moving down the assembly line in New York. The aircraft, an F-105F-1-RE (63-8366), was formally accepted by the Air Force in January 1965, and the F-105 production line was shut down. The production of the F-105F had been rapid—one was delivered in FY63, 83 in FY64, and the final 59 during FY65. The air-

An early F-105F-1-RE (62-4414) showing off its new camouflage paint scheme. The large ejection seat warning triangles would soon be replaced by much smaller and more discrete warnings. Note the BLU-1 (or -27) fire bomb shape under the centerline station. *(Dave Ostrowski via the Mick Roth Collection)*

craft cost $2,200,000 each, about $60,000 more than the F-105D. After a brief period serving as the transition trainers they were designed as, the F-model would start winning fame for combat missions over Vietnam. By early 1968, the cost per flight hour for an F-105F was still $1,020, even though maintenance hours per flight hour were down to approximately 35.[88]

F-105G (Take One)

In late 1958 Pratt & Whitney proposed a new version of the J75 with the company designation JT4B-24. The engine had the same dimensions and weight as the J75-P-19W used in the D-model but produced 30,000 lbf at sea-level, a 3,500-lbf improvement. This promised a large increase in acceleration and rate-of-climb performance, as well as an increase in service ceiling and top speed. Using the afterburner it was estimated that maximum speed would increase to Mach 2.29 at 36,000 feet and maximum sea-level rate-of-climb would be 46,000 feet per minute. A clean aircraft would accelerate from Mach 0.8 to 1.8 in 3 minutes versus the 4 it took a stock D-model to do it.

Because the D-model had included eight distinct production blocks, maintenance and logistics were more difficult than desired. In an effort to eliminate this problem and improve the reliability and serviceability of the aircraft, in 1962 Republic proposed to bring all 475 D-models produced prior to the block-31 aircraft up to a new standard known as the F-105G.[89] The design would have featured a 31-inch fuselage stretch to house a 220-gallon fuel tank, added saddle and area-rule bump tanks, and installed the Pratt & Whitney JT4B-24 engine. The landing gear would also have been strengthened, all outstanding TCTOs incorporated, and various equipment changes made to improve reliability. The cost for the program was estimated at $321,300,000, or $71,400 per aircraft. Assuming approval in July 1962, the first aircraft was to be delivered in October 1963, with all 475 modified by September 1965. Although the proposal was well received within the Air Force, all F-105 money was already earmarked for the F-105F program.

F-105H

The proposed F-105H[90] of 1964 was based on the F-105F and featured a new wing with 448 square feet of area and folding wingtips. A new stabilator with 133 square feet of area, a larger vertical stabilizer, and a retracting ventral fin were also included. The M61 Vulcan cannon and its ammunition would be relocated to an external pod and the volume in the nose used by additional avionics. The Pratt & Whitney JT4B-24 engine would replace the J75-P-19W, with additional fuel housed in the weapons bay and under the second seat.

The main landing gear was to be replaced by a tandem wheel type reminiscent of the Republic XF-91 interceptor of 1949, but retracting normally (i.e., inboard). New trailing-edge and leading-edge flaps, ailerons, spoilers, landing gear fairings, and pylons rounded out the concept. Although represented as a "modification," in reality it would have been a new aircraft. There is no record of how the Air Force viewed the proposal, but in any case it never progressed beyond the paper stage.

Other Stillborn Thuds

In 1958, the West German Luftwaffe was equipped with various versions of the F-84 and was looking for a more modern fighter to replace them. The F-105 was short-listed as a possible replacement. During an evaluation of 14 types during May 1958, the F-105 (along with the F-102, F-106, and BAC Lightning) was rejected as too expensive to buy and maintain. Nevertheless, on 18 June 1958, Republic submitted a proposal for coproduction that included setting up a second final assembly line in Europe. Similar arrangements were proposed to the Belgian and Dutch governments, at a unit cost of some $1,400,000. Half of this cost would be paid by the governments involved, the other half being financed by U.S. Military Assistance Program funds. All of these countries eventually would choose the Lockheed F-104 Starfighter.

The Air Force issued a requirement for a tactical strike-reconnaissance system designated SR-195 in September 1958. Republic's answer was the AP-63-36, a growth version of the F-105D to be built around advanced electronics, including a truly all-weather bombing system. Its reconnaissance systems would be either electronic, via the installed avionics, or photographic, via a podded camera installation fitted in the weapons bay. The design featured a new nose section with 32 inches added to accommodate a second crew member under a canopy similar to that of the stillborn E-model. The wing span was to be enlarged by 4 feet with the addition of new wingtips. Additionally, several options were presented, including boundary-layer control, saddle tanks, and the JT4B-24 engine. An economy version was also presented that deleted the new wingtips and differed in several minor details. As it wound up, SR-195 would not progress beyond the proposal stage.

In July 1960, Republic offered the F-105D to the French *Armée de l'Air* and the British Royal Air Force (RAF). Both offers included on-site production, described the performance of the block-25 aircraft from 425 different NATO airfields, and emphasized its capability to carry 13,000 pounds of armament and its ruggedness. Quoted costs, excluding government furnished electronics and other equipment, were approximately $1,350,000 per F-105D and $1,710,000 per F-105F based on a 100-unit production run. The French proposal pointed out that a version of the J75 had already been licensed for French production and was completely interchangeable with the -19W. The British were given the option of modifying the aircraft to accept the same Bristol Olympus B.01.22R used in the TSR-2 or purchasing J75-P-19Ws from Pratt & Whitney. The Olympus engine made 43,200 pounds of thrust with afterburning and water injection, and would have substantially improved the performance of the F-105. Much to Republic's disappointment, neither country accepted the proposal.

During February 1962, the Aeronautical Systems Division asked Republic if the F-105D had the capability to set a new world speed record. An engineering analysis showed that it could break several of the Class C, Group III (aeroplanes with jet engines) records, particularly the 3-kilometer restricted altitude course record that had been set by a McDonnell F4H-1 (F-4A) Phantom II. A production F-105D could almost do the job on a standard day, but a few simple modifications would boost the speed to 976 mph, and a little tweaking of the J75 could probably raise it to the magic 1,000 mark. The modifications involved the deletion of the tail hook, sealing the aircraft, and removing all protuberances. More extensive modifications including adding area-rule bumps ahead of the tail to decrease drag, and adding exit nozzle blisters similar to those on the F-102A would have guaranteed a substantial record increase. It was an interesting possibility, but for unrecorded reasons this attempt was never made.

In August 1964 Republic proposed installing an advanced avionics system in all F-105s to improve the aircraft's low-level penetration capabilities. The major changes included an inertial measurement unit, a dual-frequency multimode radar, a new automatic flight control system, a new digital computer, a laser range finder, and a heads-up display. In addition, F-models would receive a rolling map display in the rear cockpit. The company proposed a three-phase program, with the first phase consisting of a 15-month flight test program involving one F and two D-models. Phase 2 would modify one squadron of mixed Ds and Fs for operational evaluation, and the final phase was the modification of the entire F-105 fleet by December 1968. The program would have 100 F-105s out of service at any one time, but with the war in Vietnam heating up, the Air Force simply could not spare that many aircraft, and the proposal was not acted upon.

Republic made one last effort to reopen the Thunderchief production line in May 1968. Republic presented a proposal to the Department of Defense to build 300 new F-105Fs at a fly-away cost of $2,200,000 each. This price was for a trainer-equipped two-seater and did not include any of the electronics being fitted to the aircraft for Southeast Asia service. The company also presented a proposal that would modify the remaining 344 F-105D aircraft to a block-35 standard that included the Thunderstick II bombing system, a modified engine, and more internal fuel. According to the proposal, the F-105D-35-RE would have had a 20 percent payload improvement, a 27 percent range increase, a 73 percent decrease in vulnerable areas, and a maneuvering limit raised to 7-g. Fairchild Hiller, who had by now taken control of Republic, guaranteed a 13-month turnaround at a unit cost of $1,800,000. But the F-105 was getting old and too few of them remained in the inventory to justify a major modification program, so the Air Force declined, and the beginning of the end for the Thunderchief was in sight.

Notes

61. *The F-105: A Chronology (1951–1973)*, History Office, Mobile Air Materiel Area, Brookley AFB, Alabama, September 1963, pp 80–81.
62. History of the Air Force Plant Representative, Republic Aviation Corporation, Farmingdale, Long Island, New York, 1 January 1958 through 30 June 1958, p 53.
63. *Development of Airborne Armament, 1910–1961*, Historical Division, Air Force Systems Command, Aeronautical Systems Division, Wright-Patterson AFB, Ohio, October 1961, p III–447.
64. Knaack, Marcell Size, *Post-World War II Fighters, 1945–1973*, Office of Air Force History, United States Air Force, 1986, p 193.
65. *The F-105: A Chronology (1951–1973)*, History Office, Mobile Air Materiel Area, Brookley AFB, Alabama, September 1963, pp 132–144.
66. Knaack, Marcell Size, *Post-World War II Fighters, 1945–1973*, Office of Air Force History, United States Air Force, 1986, p 197.
67. *The F-105: A Chronology (1951–1973)*, History Office, Mobile Air Materiel Area, Brookley AFB, Alabama, September 1963, p 135.
68. These had previously been known as the Far East Air Forces (FEAF).
69. Knaack, Marcell Size, *Post-World War II Fighters, 1945–1973*, Office of Air Force History, United States Air Force, 1986, p 198.
70. Development of Airborne Armament, 1910–1961, Historical Division, Air Force Systems Command, Aeronautical Systems Division, Wright-Patterson AFB, Ohio, October 1961, pp III–449.
71. *The F-105: A Chronology (1951–1973)*, History Office, Mobile Air Materiel Area, Brookley AFB, Alabama, September 1963, p 172.
72. *An Analysis of the F-105 Weapons System in Out-Country Counter Air Operations*, Air War College case study, Air University, Maxwell AFB, April 1968, p 27.
73. Knaack, Marcell Size, *Post-World War II Fighters, 1945–1973*, Office of Air Force History, United States Air Force, 1986, p 199.
74. *Ibid.*, p 200.
75. F-105 System Program Report, Sacramento Air Materiel Area, 31 May 1970. Air America performed a great deal of depot-level maintenance on the F-105s (and other aircraft) based in Southeast Asia. This included replacing the wing skins and center box, overhauling the M61 cannon, and performing some rewiring.
76. *The F-105: A Chronology (1951–1973)*, History Office, Mobile Air Materiel Area, Brookley AFB, Alabama, September 1963, p 148.
77. Still the F-110 at the time.
78. Knaack, Marcell Size, *Post-World War II Fighters, 1945–1973*, Office of Air Force History, United States Air Force, 1986, p 199.
79. *The F-105: A Chronology (1951–1973)*, History Office, Mobile Air Materiel Area, Brookley AFB, Alabama, September 1963, p 95.
80. History of the Air Force Plant Representative, Republic Aviation Corporation, Farmingdale, Long Island, New York, 1 January 1958 through 30 June 1958, p 54.
81. Republic Shop Order #33911, revision G, *Conversion of "E" to "D,"* 19 November 1959.
82. Knaack, Marcell Size, *Post-World War II Fighters, 1945–1973*, Office of Air Force History, United States Air Force, 1986, p 201.

83. *Ibid.,* p 201.
84. Exactly what this had to do with the conversion of the F-105D to a two-seat aircraft is unclear. More probably it was a deficiency noted when the last D-models incorporated the dual in-flight refueling capability and was scheduled to be corrected in the next production block, which just happened to be the F-105F.
85. *Introduction to the F-105F Two-Place Airplane,* Republic Aviation Corporation, undated (probably early 1963), p 1.
86. *The F-105: A Chronology (1951–1973),* History Office, Mobile Air Materiel Area, Brookley AFB, Alabama, September 1963, p 171.
87. *Introduction to the F-105F Two-Place Airplane,* Republic Aviation Corporation, undated (probably early 1963), p 12.
88. Knaack, Marcell Size, *Post-World War II Fighters, 1945–1973,* Office of Air Force History, United States Air Force, 1986, p 203.
89. An unofficial Republic marketing designation.
90. Another unofficial Republic marketing designation.

F-105D/F Operational Service—Vietnam and Thud Ridge

On 2 August 1964, the North Vietnamese attacked the destroyer USS *Maddox* (DD-731) in international waters in the Gulf of Tonkin. Two nights later, *Maddox* and the USS *Turner Joy* (DD-951) were attacked.[91] Even at the time, there was a debate over whether the second attack actually took place, with the North Vietnamese denying it. It did not really matter. The politicians in Washington had already made up their minds, and on 7 August Congress passed the Gulf of Tonkin Resolution, abdicating its war-making powers to the president. President Lyndon B. Johnson promptly committed forces en masse into Vietnam. In short order, two B-57 squadrons moved from Clark AB in the Philippines to Bien Hoa, and F-100 and F-102 fighter squadrons arrived at Da Nang. On 9 August, eight F-105Ds from the 36th TFS at Itazuke, Japan, arrived at the Royal Thai Air Force Base (RTAFB) at Korat, Thailand. It was the beginning of a long eight years.

The F-105's combat debut came on 14 August 1964. A flight of four F-105Ds from the 36th TFS participated in a rescue combat air patrol (RESCAP) by attacking an antiaircraft site in the Plaine des Jarres. The F-105D (62-4371) piloted by Lt. Dave Graben was severely damaged by Pathet Lao 37-mm AAA and started burning. Graben managed to get the fire out and return to Korat. This was the first of a long series of battle-damaged Thunderchiefs that would bring their pilots home. The F-105s attacked minor targets inside Laos for the remainder of 1964 while everybody hoped that a permanent settlement could be found to the crisis.[92]

With twice the bomb load and half again as much speed as the F-100D, the F-105D would deliver over three-quarters of the bombs dropped against North Vietnam during the first three years of the war. The first offensive mission was flown on 13 January 1965 when 16 aircraft from the 44th TFS and 67th TFS destroyed a bridge at Ben Ken in northern Laos with M117 750-pound general-purpose bombs and AGM-12B Bullpup missiles.

During early 1965 there were approximately 150 F-105s assigned to PACAF, and they were deployed TDY to Korat, followed in early 1965 by deployments to Takhli RTAFB. These two bases would be used by the F-105s for the rest of the war, with personnel from almost every F-105 squadron showing up at some point. An exception to using the two Thai bases was the 23rd TFW, which sent the 562nd and 563rd TFS to Son Nhut in June 1964 for eight

An F-105D-25-RE (61-0169) from the 563rd TFS on the ramp at Da Nang in May 1965. The silver paint applied during Project Look Alike is already showing signs of wear due to the weather conditions in Southeast Asia and the daily combat sorties. *(David Menard via the Larry Davis Collection)*

months. Both Thai bases had provisional wings made up of TDY personnel that rotated in and out on a timely basis. Korat was home for the 6234th TFW(Provisional), while the 6235th TFW(P) was at Takhli.

On 7 February 1965, the Viet Cong attacked a U.S. base camp in South Vietnam. The United States responded with Operation Flaming Dart, which authorized attacks inside North Vietnam for the first time. On 8 February F-105s attacked the Dong Hoi Army Complex in retaliation for the Viet Cong attack. Two days later, the Viet Cong attacked the U.S. base camp at Qui Nhon, killing 23 servicemen. The U.S. response was Operation Flaming Dart II, a series of coordinated attacks against various bases in the panhandle region of North Vietnam.

During February 1965, while Flaming Dart II was ongoing, a list of targets was drawn up by the Joint Chiefs of Staff in the Pentagon and approved by the White House. These included mainly military targets—POL depots, ammunition storage, military barracks, and bridges used to transport military supplies. Industrial and dual-use targets (power plants, etc.) were specifically forbidden. But the North Vietnamese and Viet Cong continued their attacks. On 2 March 1965 President Johnson ordered the beginning of Operation Rolling Thunder to strike the dual-use targets, and the F-105s would fly the majority of the missions. Rolling Thunder was initially targeted toward radar installations and bridges between the DMZ on the 17th parallel, and the 19th parallel just south of Hanoi. After an abortive attempt at a cease-fire in mid-May, the operation was extended to the 20th parallel.

Flying from Korat, 25 F-105s launched an attack on 2 March 1965, losing two aircraft to AAA fire. The next day a mixed force of 44 F-105s, 40 F-100s, 7 RF-101s, and 20 B-57s attacked the Xom Bong ammunition dump. On 3 April 1965, 46 F-105Ds, supported by 21 F-100Ds from South Vietnam, struck the Thanh Hoa Bridge, and another 48 F-105Ds returned the next day. Four North Vietnamese MiG-17s shot down two of the F-105s, and a third was downed by AAA fire. As for the bridge, it survived the Bullpup missiles and 750-pound bombs and remained in operation until 1972.

Three F-105Ds (60-0429, 61-0048, and 58-1163) from the 355th TFW begin refueling prior to a strike against North Vietnam under Operation Skypoint I in November 1965. All three aircraft carry six Mk 82 500-pound general-purpose bombs with Snakeye retarding fins and a pair of Mk 82s under the wings. The target is probably along the Ho Chi Minh Trail, requiring a low-level attack. The furthest aircraft (58-1163) shows that it has already been retrofitted with the boom-style refueling receptacle. *(U.S. Air Force)*

Later Rolling Thunder missions were usually made up of 16 F-105s in four elements of four aircraft each. These were covered by two flights of four fighters, generally F-4Cs, while EB-66 electronic warfare aircraft jammed North Vietnamese communications from safer areas off the coast. During Rolling Thunder, F-105s averaged 272 sorties per week.

North Vietnam was divided into six zones known as "Route Packages," numbered one to six heading north from the DMZ. The higher the number, the heavier the defenses. Route Pack 6 included the areas around the main port city at Haiphong and the capital city of Hanoi. Route Pack 6A covered the area directly around Hanoi. Naturally, this is where the North Vietnamese concentrated a high percentage of the air defenses. Eventually Hanoi would be ringed with thousands of antiaircraft guns and several hundred SA-2 surface-to-air missile sites.

Strikes against targets near Hanoi involved 1,250-mile round trips from Takhli. The high ambient temperatures which usually existed in Thailand degraded takeoff performance, which often required departures with less than a full fuel load, while the climb to altitude in the hot, thin air resulted in higher fuel consumption. Frequently the Thunderchiefs departed with almost no fuel and immediately refueled heading north. A thousand miles later they again refueled over Laos before crossing into North Vietnam. This refueling operation often had to be repeated on the way back, especially if afterburners had been used to evade enemy defenses. On occasion, KC-135 tankers, very much against orders, would penetrate North Vietnamese airspace to rescue F-105s short on fuel or suffering from bat-

A pair of 355th TFW Thunderchiefs en route to targets in South Vietnam during May 1970. Both the F-105D from the 44th TFS, and the F-105F-1-RE (63-8285) from the 333rd TFS, are armed with eight Mk 82 500-pound bombs. *(U.S. Air Force)*

Mechanics from the 388th TFW run the engine up during night maintenance activities at Korat in 1966. The F-105D was a very complicated airplane and required between 35 and 100 hours of maintenance for each hour of combat flight time. The APR-25 RHAW antenna housing is located under the fuselage just behind the radome, although this F-105D-31-RE (62-4334) is not equipped with a strike camera. *(U.S. Air Force)*

tle damage. Many F-105 pilots escaped being guests in the "Hanoi Hilton" because of the courage and skill of the KC-135 crews.

Hanoi was located in the Red River valley. There was a small mountain range on the north side of the river just before the river widened out into the flats surrounding the city. On the other side of the mountains was the Black River. When approaching Hanoi from Thailand, the F-105s had to cross these mountains. The route into, and out of, Route Pack Six had been largely dictated by politicians in Washington. Flying the same route every day gave the North Vietnamese a chance to significantly strengthen the defenses along the mountains. Soon, so many F-105s had been shot down over these mountains that they became known as "Thud Ridge."

Once over Thud Ridge, the F-105s would approach their targets low and fast, an environment in which the F-105 excelled. Maneuverability and stability during high-speed, low-level flight were excellent due to the aircraft's high wing loading. They were usually loaded with a single MER on the centerline with up to six M117 750-pound bombs and two 450-gallon wing tanks. Often the bombs carried fuze extenders to ensure they exploded slightly above the ground, maximizing damage against soft targets. For harder targets, the F-105s carried 2,000-pound bombs under the wings, with a 600-gallon drop tank under the fuselage. Heavily fortified targets, such as the Paul Doumer Bridge, required M118 3,000-pound bombs, which were also carried under the wings. Early in the war, the F-105s often carried AGM-12B Bullpups, but these missiles were difficult to use since the pilot had to provide guidance all the way to the target, leaving him exposed to ground fire. It was not until later that ECM pods and AIM-9s became standard equipment. Although the AGM-45 Shrike had been baselined for the F-105 as early as 1963, it was not until 1966 that this missile began to be used to kill SAM radars, mainly in the hands of the Wild Weasels.

The run down the Red River valley was made at 600 knots and 1,500 feet. As they closed in on the target the pilots would engage the afterburner and pitch up to 3,000 feet or so, then bank over and attack. Assuming they made it through the defenses around the target, the pilots left the afterburner engaged and departed the area as quickly as possible.

This F-105D-10-RE (60-0504) from the 357th TFS carries a famous name—"Memphis Belle II." Note the M118 3,000-pound bomb equipped with a fuze extender under the wing. *(Paul Minert Collection via Mick Roth)*

This F-105D -6-RE (60-0409), named "Dixie Twister," sits in one of the revetments at Korat that were built in November 1966. Six M117 750-pound bombs are on the centerline MER. This aircraft also carries the APR-25 antenna fairing under the nose but does not have the strike camera installed. *(U.S. Air Force)*

And the North Vietnamese defenses were good. During 1965, when the defenses were just getting organized, 59 Thunderchiefs were lost to enemy fire. Things got worse in 1966 when the two wings together lost 103 F-105Ds and 6 F-105Fs. So many F-105s were going down that the Air Force hurried the conversion of the two USAFE wings to F-4Cs to free up their F-105s for the combat squadrons. The statistics were not much better for 1967, when F-105 losses included 84 D-models and 12 F-105Fs.[93]

When it became obvious that the war was going to last longer than expected, PACAF set up the 355th TFW at Takhli on 8 November 1965 and the 388th TFW at Korat on 8 April 1966 to replace the provisional command structure. For the next two years the F-105s from these two wings would fly almost daily combat sorties against North Vietnam. The TDY squadron personnel rotated back to their home bases, but the aircraft remained behind, flying thousands of hours in the hostile environment. Eventually, all F-105D/Fs came to Southeast Asia. Only the F-105Bs were spared, mainly because they were not considered combat capable since they lacked the ASG-19 fire control system (the MA-8 never did prove effective). The F-105s in USAFE were being replaced by the F-4D, and the last Thunderchiefs departed Europe in March 1967, destined for Southeast Asia.

In February 1966, the 334th TFS left Thailand and returned to Seymour-Johnson after spending six months in Southeast Asia. The unit had flown 2,204 operational sorties for a total of 5,488 combat hours. They had dropped 19,840,000 pounds of bombs, fired 17,031 2.75-inch rockets, and expended 351,547 rounds of 20-mm ammunition. An average of 90 combat missions had been flown by 24 of the 27 pilots in the squadron, 3 being shot down with 2 subsequently rescued. The F-105D had demonstrated it was an excellent fighter-bomber, something of a surprise given its previous troubled history.

The F-105D was somewhat less successful as a fighter, often being hard pressed by enemy MiG-17 and MiG-21 fighters. It had a wing loading that was much too high for it to be able to maneuver effectively against the more nimble MiGs. Since all the ordnance was

This F-105D-10-RE (60-0494) shows one of the early variations in serial number presentation. It would take several years before all the units standardized how serial numbers were displayed. Armament includes six 750-pound bombs on the centerline and an AIM-9 Sidewinder under the right wing. *(Larsen/Remington via the Mick Roth Collection)*

The ground crew reloads 20-mm ammunition in an F-105D-31-RE (62-4387) from the 354th TFS during December 1969. The ammunition drum was located high above the ramp, necessitating this interesting rig attached to a standard bomb truck. *(Cradle of Aviation Museum via Ken Neubeck)*

A J75 engine being installed into an F-105D in December 1968. As in many Century Series fighters, the entire rear fuselage had to be removed to gain total access to the engine. Like the F-105 itself, the J75 was very large and required a great deal of effort to move, even on the wheeled engine stand. Early in the F-105's career, the J75 needed overhaul every 300 hours, but by 1961 the interval had doubled to 600 hours. *(Cradle of Aviation Museum via Ken Neubeck).*

carried externally, maximum performance could only be attained once the bombs and rockets had been jettisoned. However, when jumped by MiGs, the thrust of the J75 enabled a cleaned-up Thunderchief to go supersonic "on the deck," quickly leaving its pursuers behind. When forced to fight, the F-105 did manage to shoot down 27.5 MiG-17s during the course of the war, all but 2 with the 20-mm cannon. The other two were downed by Sidewinders, and the half-kill was shared with an F-4 crew. Captain Max C. Brestel, an F-105D pilot of the 355th TFW became the first double MiG-killer of the war when he shot down two MiG-17s during a 10 March mission.[94] One F-105F (63-8320) is unofficially credited (with three different crews) with the destruction of three MiGs—one by 20-mm gun fire, one with an AIM-9, and the last was downed when the F-105F jettisoned his centerline MER full of bombs directly into the path of the MiG. In return, North Vietnamese MiGs shot down 21 F-105D/Fs between 1965 and 1972.

Perhaps the most famous F-105 raid occurred on 2 August 1967, when they attacked and heavily damaged the Paul Doumer bridge north of Hanoi. This bridge was vital for moving war supplies between China and North Vietnam, and was a prime target. However, the bridge was eventually repaired. Nevertheless, it was hit and knocked down again—several times.

Two things happened that would cut the 1968 losses drastically. First, President Johnson ordered that all bombing would be restricted to Route Pack 1 beginning on 31 March, then completely halted all bombing of North Vietnam on 1 November. And at the same time, the Air Force began reequipping the 388th TFW with F-4E Phantom IIs. The only F-105s going north would be the Wild Weasels. Total combat losses for 1968 were 28 F-105Ds and 6 F-105Fs.[95]

First Lieutenant David B. Waldrop from the 34th TFS shot down this MiG-17 on 23 August 1967 using the 20-mm cannon. Although the F-105 was not noted as a dogfighter, when forced to fight, the F-105 did manage to shoot down 27.5 MiG-17s during the course of the war, all but 2 with the 20-mm cannon. The other two were downed by Sidewinders, and the half-kill was shared with an F-4 crew. *(Cradle of Aviation Museum via Ken Neubeck)*

Although the air war into North Vietnam had, for all intents and purposes, ceased for the F-105 strike crews, the combat over Laos and Cambodia continued. Attacks against the Ho Chi Minh Trail were tripled, and the North Vietnamese brought in large numbers of antiaircraft guns to protect their vital supply routes. From 1969 until the Weasels left in 1974, F-105 losses amounted to 18 D-models, 2 F-105Fs, and 11 G-models.

"The Mercenary" was an F-105D-10-RE (60-0512), shown here carrying six M117 750-pound bombs and an AIM-9 Sidewinder. This aircraft would be shot down over Laos on 1 September 1968 with Captain D. K. Thaete at the controls. Thaete was injured in the incident, although he was subsequently rescued and returned to the United States. *(Larsen/Remington via the Mick Roth Collection)*

In late 1968, the 388th TFW relinquished the last of their F-105s to the 355th TFW at Takhli. Although the F-105s were no longer flying over North Vietnam, they were still assigned targets in South Vietnam, and during August 1970 the 355th began attacking targets in Cambodia. In general the F-105s used M117 750-pound bombs, or the newer Mk 82 500-pound and Mk 83 1,000-pound bombs, with a few M118 3,000-pound bombs and

"Sittin Pretty" was an F-105D-15-RE (61-0078) that carried its serial number in two locations on the vertical stabilizer—once in large white numbers, and once in smaller black numbers. This aircraft was shot down over North Vietnam by 37mm AAA on 3 September 1967. Captain H. W. Moore, Jr. was listed as MIA. *(Larsen/Remington via the Mick Roth Collection)*

Yet another variation in serial number presentation is shown on this F-105D-10-RE (60-0445). Note the black USAF and the white serial number. Other variations included both in white and both in black. The small B on the tail was an early tail code used by the 44th TFS during 1967. Six M117 750-pound bombs are on the centerline, and an LAU-3 rocket pod is under the right wing. *(Larsen/Remington via the Mick Roth Collection)*

Before the two-letter tail codes became standardized, some units used tail codes of their own making, with "P" being used by the "The Great Pumpkin" (and, apparently, this is what the P stood for). This F-105D-10-RE (60-0421) carried a cartoon Snoopy "the World War I Ace" sitting on his doghouse on the fuselage. *(Larsen/Remington via the Mick Roth Collection)*

"Old Crow II" was flown by Colonel Clarence E. "Bud" Anderson, Commander of the 355th TFW at Takhli. Subtle colored stripes just behind the radome, around the pilot, and on the radar reflector on the nose gear strut represent each of the four squadrons assigned to the Wing. The Presidential Unit Citation and Air Force Outstanding Unit ribbons are on the fuselage under the windscreen. This F-105D-10-RE (60-5375) was actually assigned to the 333rd TFS, evidenced by the RK tail code. This aircraft was later modified to the T-Stick II configuration by Air America at Tainan, Taiwan. *(C. E. Anderson)*

CBU-24 cluster bombs also being used. When flying against high-threat targets, the F-105s generally released their bombs high enough to ensure a recovery altitude of 4,000 feet. Against low-threat targets this was allowed to drop as low as 1,500 feet.[96]

Interestingly, AGM-12C Bullpups were still being used in 1970. Planners used an expected reliability of only 65 percent when assigning the missiles to specific missions. However, during the six-month period ending in June 1970 the 355th TFW experienced an actual reliability of 80.77 percent. This was apparently highly variable, and the following three months showed only a 68.6 percent reliability. The missiles were primarily used against caves, although some were also fired at roads, bridges, and trucks. Bullpups were normally released in a 20° dive at about 18,000 feet slant-range, or somewhat closer. The AGM-12Cs being used at the time had a slightly shorter than normal slant range because they were equipped with AGM-12B nose cones.[97]

Like the F-4, many F-105s were equipped with a bomb damage assessment (BDA) camera, in this case mounted under the nose just behind the radome. Strikes were reviewed upon return to Takhli or Korat. But for the cameras to record anything useful, the pilot had to stay wings-level and directly overfly the target. Needless to say, pilots were less interested in taking useful photos than escaping with their lives, leading to only seven percent of the BDA film being useful in actual bomb damage assessments. Most of the time, except over North Vietnam, the actual assessments were made by the forward air controller (FAC) that was directing the bombing at the target. Over the north, dedicated reconnaissance aircraft or drones surveyed the damage.[98]

This F-105D-10-RE (60-5375) shows the installation of the strike camera behind the APR-25 antenna fairing under the nose. Not all aircraft were modified with the strike camera, which proved to be of marginal value in combat since it forced the pilot to directly overfly the target "wings-level"— something most pilots were unwilling to do for risk of being shot down. *(C. E. Anderson)*

"Bubbles I" was an F-105D-5-RE (58-1157) based at Korat RTAFB. Note the gray-painted fairings for the APR-25 antennas and preamplifiers on the vertical stabilizer. Later these would be camouflaged like the rest of the aircraft. *(Larsen/Remington via the Mick Roth Collection)*

The F-105 was still proving to be an effective combat aircraft. Between July–October 1970, the 355th TFW scheduled 2,904 combat sorties, and 2,886 of these were actually flown—a 99.4 percent dispatch rate. These encompassed 7,349 combat flying hours, and resulted in the loss of one F-105D (61-0153), but fortunately the pilot ejected successfully and was rescued. But this was not the whole story. In the period from 1 December 1969 to 6 October 1970, only 77 percent of the average 66 Thunderchiefs assigned during the period were operationally ready. Weapons used during this 10-month period included 1,176,382 rounds of 20-mm ammunition, 13,747 M117 bombs, 733 M118 bombs, 60,760 Mk 82 bombs, 1,152 Mk 84 bombs, 405 AGM-12C Bullpups, 2,976 CBU-24 cluster bombs, and 11 AGM-45 Shrikes.[99]

Generally, pilots and EWOs got to rotate home after 100 combat missions. Three pilots, however, stayed and racked up 200 combat missions each. Major Larry Waller flew 100 missions with the 355th TFW and then flew a second 100 missions when assigned to the 388th TFW. First Lieutenant Karl Richter flew at least 230 missions, although not all of them were logged in official records. Captain Peter Foley from the 469th TFS was the third pilot to complete 200 missions.

During the time the 355th TFW was in Southeast Asia, they flew 101,304 combat sorties for a total of 263,650 hours. The Wing hit 12,675 targets, dropping 202,596 pounds of ordnance. On 9 October 1970, four Thunderchiefs from the 334th TFS of the 355th TFW flew the last F-105D combat mission of the war. The last F-105D left Takhli on 10 November 1970, making the long trip back to the United States, but the F-105F/G Wild Weasels were transferred to Korat on 21 September 1970 and assigned to the 388th TFW. The Weasels, originally just Det 1 of the 12th TFS, became the 6010th Wild Weasel Squadron, then the 17th Wild Weasel Squadron. They remained at Korat throughout the remainder of the war, including both Linebacker I and II operations in 1972. They were the last F-105 squadron to come home, in November 1974.

By the end of the war, the F-105D fleet had been decimated. Of the 610 F-105Ds that were built, 342 had been lost in Southeast Asia. Of the 143 F-105Fs built, losses in Southeast Asia amounted to 36 aircraft, including at least 15 of the sophisticated F-105Gs. And that doesn't count any operational losses caused by "normal" accidents in other locations the F-105 had operated (Europe, the United States, etc.). Most of the combat losses came at the hands of North Vietnamese antiaircraft guns—known SAM kills accounted for only 32 Thunderchiefs.[100]

SEA Modifications

Use of the F-105D and F in Southeast Asia necessitated numerous modifications for problems created by the constant high-speed, high-g, low-level combat flying. The aircraft had always suffered from high temperatures in the aft fuselage near the afterburner, some of which had been alleviated by the inclusion of the small scoop at the base of the vertical stabilizer and two small vents on the aft fuselage added to the F-105D. But these proved inadequate in the heat and humidity of Thailand. Republic designed a pair of small air scoops that fit over the original vent locations to bring cool air into the aft fuselage. As with the engine air intakes, these scoops had small bleed openings that carried away excess amounts of air that built up inside the scoop.[101]

Beginning in mid-1965, the F-105s were given a camouflage paint job to replace the Look Alike silver. The colors were two dark greens and a tan on the upper surfaces, which made the F-105s harder to see from above against the Vietnamese jungle terrain. The undersides were painted a light gray. The camouflage paint also resealed the many fuselage panels, which were subject to moisture invasion in the humidity of Southeast Asia. Personal markings quickly became standard, even though they were officially banned by PACAF policy.[102] Generally the markings appeared on the fuselage just under the air intakes (often with a

Beginning in 1966, F-105s started wearing the standard Vietnam-era camouflage of tan (FS-30219) and two shades of green (FS-34079 and FS-34102) on top, with light gray (FS-36622) on bottom. Approximately one aircraft in every four had the tan and 30219 green reversed, as shown on this F-105D-5-RE (59-1750). *(Larsen/Remington via the Mick Roth Collection)*

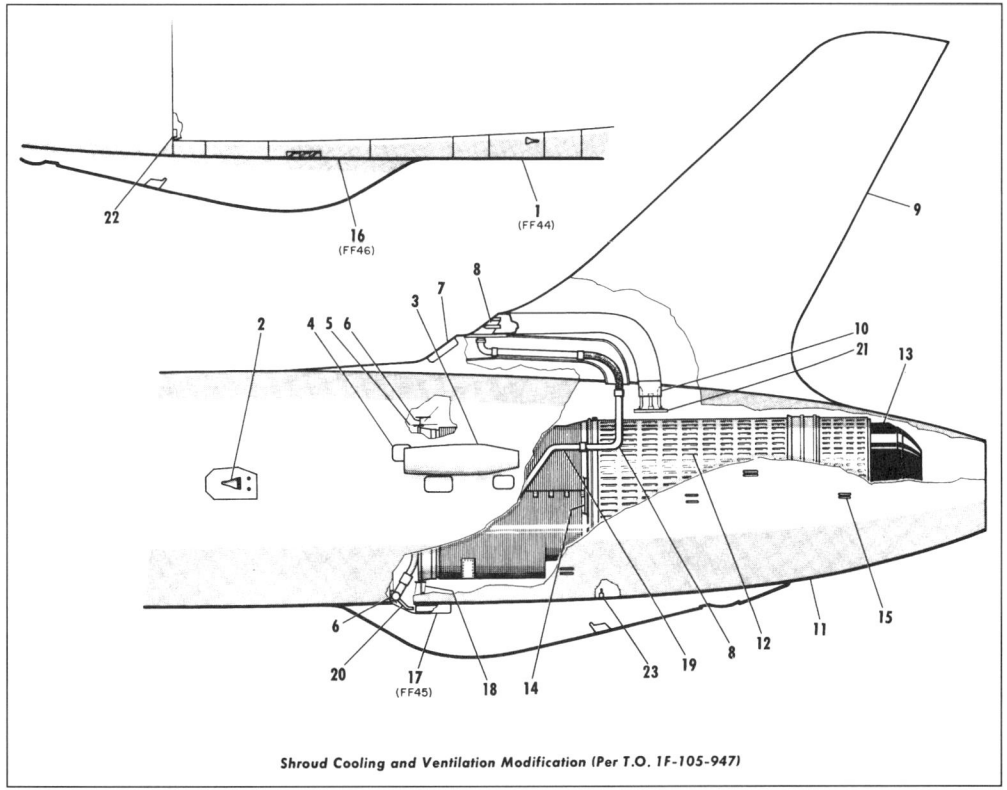

The F-105 had always suffered from high temperatures in the aft fuselage near the afterburner. Republic designed a pair of small air scoops that fit over the original vent locations to bring cool air into the aft fuselage. The ducting from the scoops at the base of the vertical stabilizer was also modified to more efficiently cool the afterburner. *(U.S. Air Force)*

Another aircraft painted in "reverse" camouflage was this F-105D-5-RE (58-1167) named "Miss Universe." Note the particularly small serial numbers on the vertical stabilizer and the strike camera mounted under the nose just behind the APR-25 antenna fairing. *(Larsen/Remington via the Mick Roth Collection)*

name on the side of the intake itself). Particularly on the Weasels, shark mouths appeared on the nose (although never looking quite as good as they did on the F-4Es).

The F-105D/F proved to be able to withstand serious battle damage, bringing crews home time and again with extensive damage. One aircraft was hit by a SAM in the rear fuselage, resulting in a broken main fuselage frame, 87 holes, and the loss of major portions of the vertical stabilizer and rudder. The pilot was wounded in the left hand and leg. But the Thunderchief kept flying, even going through a relatively routine in-flight refueling, before putting down safely at Udorn, the nearest friendly air base.

There were some areas where damage produced very different results, however. When Republic had designed the F-105, conventional battle damage was not foremost in their minds. The aircraft had redundant hydraulic systems, but primarily to guard against a hydraulic failure, not damage to one system. As a result, the hydraulic lines from both systems ran next to each other for the entire length of the fuselage. A single hit in this area was likely to disable both systems, resulting in complete loss of flight controls. Loss of both hydraulic systems usually locked the stabilators in the up position, which put the aircraft into a nose-down attitude—straight down into the jungle below.[103]

Republic came up with a quick fix for the stabilator problem by installing a simple mechanical lock on the stabilator. When the pilot lost his hydraulic systems, he could mechanically lock the stabilators in a neutral position and use the ailerons during the egress out of North Vietnam. Later, Republic came up with a third hydraulic pitch control system that gave the pilot some control even if the primary and secondary systems were shot away. The hydraulic lines for this system ran along the top of the fuselage, covered by a small hump running between the rear of the canopy fairing and the vertical stabilizer.

An immediate requirement existed to extend the UHF range of Rolling Thunder strikes to ensure continuous reliable communications with the Da Nang CRC (control and reporting center). To overcome inherent line-of-sight limitations to UHF radio communications,

The three different paint schemes worn by F-105Ds in Southeast Asia show up well here. At the top are two aircraft wearing the standard camouflage pattern, while the second aircraft from the bottom wears the "reverse" scheme. The bottom aircraft is still in the Look Alike silver lacquer. These Thunderchiefs are dropping M118 3,000-pound bombs during a "buddy bombing" strike led by an EB-66 Destroyer. *(U.S. Air Force/DVIC)*

the Air Force developed a concept for automatic radio relay that permitted the reception and instantaneous retransmission of UHF signals without delay or reiteration by a relay operator. Realizing the most likely platform for the system was a fighter, and that fighters seldom have any unused internal space for more equipment, the Air Force decided to package the system in a pod.[104]

The Collins Airborne Automatic UHF Relay system was developed for use on F-105s and F-4s. The Air Force wanted a total of 16 aircraft modified to operate 12 pods. By February 1966 the Collins pod had been successfully flight tested at speeds up to 0.96 Mach, and the

This F-105D-10-RE (60-0454) flown by Maj. William McClelland received major wing damage while bombing a bridge over the Kuih Bich Dowg river in Vietnam on 7 July 1966. McClelland flew over 500 miles to get back to base, and made one in-flight refueling along the way. Just short of a year later, on 5 July 1967, Captain W. V. Frederick would be shot down in this aircraft over North Vietnam and listed as MIA. *(Cradle of Aviation Museum via Ken Neubeck)*

Admiral Radio Company was providing 16 AN/ARC-51-BX UHF radios in March and 16 more in April. The pod contained two radios (the rest were procured as spares), each tuned to a different frequency to allow reception on one radio and near-instantaneous retransmission on the other. As a secondary function, the system could also be used by the pilot as a back-up communication system in lieu of the ARC-70 command radio. The relay system consists of a radio relay pod carried on the centerline pylon, a control head, and a control panel, both mounted on the right-hand console. The pod had several rows of antennae stubs protruding out of a long, cylindrical body. At the time the pod was considered highly classified, and pilots were warned that it could not be jettisoned even in an emergency.[105]

Electronic Countermeasures

A significant modification in response to sorties into North Vietnam was the installation of a radar homing and warning (RHAW) system, which alerted the pilot to nearby AAA and SAM radars. As early as April 1965, reconnaissance aircraft had detected the first signs of SA-2 site construction in North Vietnam. On 23 July an RB-66C intercepted and recorded tracking radar signals from one of the several SA-2 sites ringing Hanoi. The next day a site northwest of Hanoi launched a pair of SA-2s at a flight of four F-4Cs, downing one and damaging the other three.

The SA-2 system normally consisted of a Fan Song radar van, several power vans, and six Guideline missile launchers, usually arranged in a modified Star-of-David arrangement with the radar van in the middle and missiles located at the points of the star. The system

This F-105D-10-RE (60-5376) from the 388th TFW took a hit from an "Atoll" air-to-air missile on 20 December 1967. Part of the missile was still embedded in the fuselage when the aircraft landed at Korat. This aircraft would survive the war and become one of the 30 T-Stick II aircraft. *(Cradle of Aviation Museum via Ken Neubeck)*

More battle damage, this time on Jack Broughton's F-105D-31-RE (62-4338) on 24 July 1967. This hole was the result of a 37-mm AAA hit. On 2 September 1967 Major W. G. Bennett would be killed when this aircraft was shot down over North Vietnam. *(Cradle of Aviation Museum via Ken Neubeck)*

was mobile and could be packed and on the road in several hours. It generally took about six hours to set the system up at a preprepared location. Two versions of the SA-2 system were used extensively in Vietnam—Fan Song B and F. The two-stage (booster and sustainer) Guideline missile was 37 feet long and weighed 4,500 pounds.[106]

It generally took the North Vietnamese about 40 seconds after they began tracking an aircraft before they launched a missile. This was not an absolute, however, and many missiles were launched in as little as 10 seconds. The missile had a minimum range of 5 miles and a maximum range of 17 miles, and the system was capable of launching 3 missiles at 6-second intervals against a single target. The missile had no self-contained guidance and had to receive guidance signals from the Fan Song radar all the way to intercept. A 400-pound warhead was surrounded by 3,600 shrapnel fragments. A proximity fuze detonated the warhead approximately 150 feet from its target, with the fragmentation pattern perpendicular to the missile's flight path. Due to the forward velocity of the missile, the fragments spread in a cone-shaped pattern and could be lethal within a several-hundred-foot radius of the detonation point. The warhead could also be detonated via a command from the ground.[107]

The other significant threat was the Fire Can radar used with 85-mm and 57-mm antiaircraft guns. The 85-mm required each projectile to be fuzed prior to firing, and this took approximately 6 seconds from the time the Fire Can radar established the target altitude. The 57-mm gun did not suffer from this limitation since it used a proximity fuze. The more common 37-mm AAA was optically sighted but was frequently collocated with the larger weapons or SAMs.

After the SA-2's combat debut, the Air Force and Navy formed various task groups to evaluate the SAM threat and determine the best and quickest methods to counter it. In August 1965 an anti-SAM seminar was convened at Eglin AFB to examine available and proposed countermeasures equipment, devise employment concepts and tactics, and recommend both short- and long-term solutions. The recommendations included the development and deployment of specially equipped anti-SAM aircraft (Wild Weasels, for the Air Force), the development of new weapons (to supplement the AGM-45 Shrike antiradiation missile), and the installation of jamming and RHAW equipment on tactical aircraft.

Under a top-priority program, the APR-25 and APR-26 RHAW systems, which were originally tested in the F-100F Wild Weasel, were installed on all F-105D/Fs operating in Southeast Asia. The APR-25 antennas were mounted on the front of a small fairing under the nose, which also housed the rear-looking strike camera on aircraft so equipped, and on the trailing edge of the fin cap. A flat fairing on both sides of the fin cap housed the aft APR-25 preamps. The APR-25 used a small cathode ray tube (CRT) display that was installed on the upper right side of the main instrument panel and a threat display unit (TDU) mounted at the top of the windscreen bow. Radial strobes on the CRT indicated relative bearing and signal strength of a threat through the use of rings on the scope. A "three-ringer" represented imminent danger. The APR-26 monitored the SAM missile guidance radar frequency, which increased in strength immediately prior to launch. A light on the TDU indicated a launch. At that point, all pilots in that area started looking for the telltale signs of a SAM launch. If the SAM could be seen in time, a pilot had a good chance of defeating it through some quick maneuvering.[108]

As early as 1957, a jamming pod mounted under the wings of fighter-bombers instead of inside the already crowded airframe had been investigated. The pod could be removed for strikes where radar-controlled weapons were not anticipated and replaced by an equal weight of munitions or fuel. This "universal" self-protection pod contained only two transmitters, limiting coverage to a narrow frequency range. Consequently, to compensate, strike planners had to have precise intelligence on enemy radar. The pod was subsequently designated ALQ-31.[109]

By the end of 1963, the Air Force was testing a family of five QRC-160 self-protection pods, one of them designed to jam Fan Song. The pods were 100 inches long, 10 inches in diameter, and weighed 175 to 200 pounds. Each pod contained a propeller-driven generator to provide electric power. The first pod used in Southeast Asia, the QRC-160-1, was largely a failure. The pod's construction could not withstand the stresses created by high-speed flight on the RF-101s that carried it, and the external shape generated aerodynamic forces that twisted the aircraft's wings. Consequently, after a brief trial in early 1965, the pod was withdrawn from service.

In January 1966, the Air Proving Ground Center at Eglin AFB began testing the improved QRC-160A. This modified pod, although structurally stronger and somewhat heavier than the original, had undergone no radical changes. Like its predecessor, it was a self-contained barrage noise jammer with four 75-watt magnetron transmitters (soon replaced by 100-watt models). Generally, one of the transmitters was tuned to counter the Fire Can, one to counter the Fan Song elevation frequency, and one to counter the Fan Song azimuth frequency. The fourth transmitter was equipped with a modulator to confuse the Fan Song radar operator. The pods radiated energy 360° around the aircraft, but only 5° above and 25° below the horizon. The new pods were 90 inches long, 10 inches in diameter, and weighed 200 pounds. Like the older model, the improved pod was simple to operate. Prior to takeoff, ground crewmen adjusted the controls on the outside of the pod to establish the center frequency and bandwidth that would jam particular kinds of radar. The pilot needed only to turn the transmitters on and off.[110]

By the end of January 1966 preliminary evaluations indicated that the QRC-160A would prove rugged enough for combat and was effective against Fan Song and Fire Can. By midyear the test results convinced General William Momyer, who had recently assumed command of the 7th Air Force, that the new self-protection pod offered a means of reducing losses over North Vietnam. He therefore requested a combat evaluation, and 25 of the pods, along with the technicians to maintain them, were sent to Thailand.

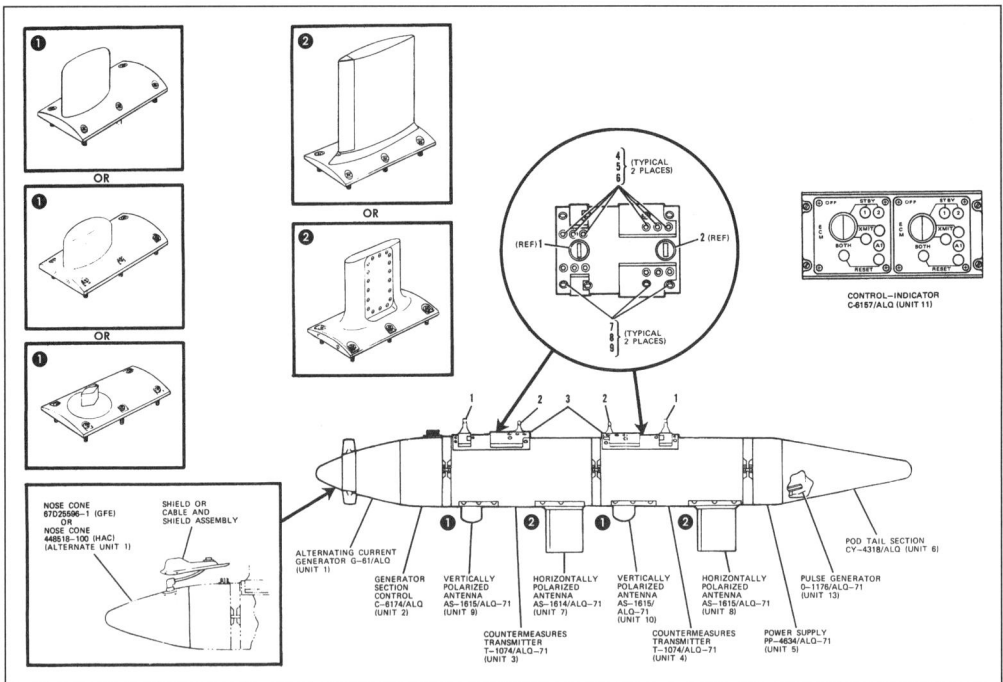

The first ECM pod to be routinely carried by the F-105 was the QRC-160A/ALQ-71. There were several variations of this pod available to jam different frequencies as needed. This required that you knew before you took off on a mission what coverage you would require. Generally, at least one ALQ-71 pod would be carried on the outer wing pylon. *(U.S. Air Force)*

By June 1966 some F-105Ds in Southeast Asia had been modified by T.O. 1F-105-890 to carry the ECM pods. The most obvious (to the crews, anyway) evidence of these modifications were the capped cannon plugs terminating in the main gear wheel wells labeled QRC-160. However, the operational squadrons discounted the pod as "high drag, heavy weight, and zero effectiveness," based largely on the unhappy results of the original QRC-160 trials on the RF-101.[111]

In September, the test team and its equipment reached the 355th TFW at Takhli and began Project Vampyrus to determine the effectiveness of the QRC-160A against radar-controlled weapons. From 26 September through 8 October 1966, F-105Ds of the 355th TFW flew 19 four-aircraft missions, sometimes as many as three a day. The project used EB-66Cs to measure the effectiveness of the pods against targets defended by SAMs and radar-controlled AAA. The electronic warfare officers aboard the EB-66Cs first verified the number, location, type, and transmission characteristics of radars protecting the targets, then observed how these transmitters reacted to the pod-generated jamming. When necessary, the EB-66Cs jammed radars outside the frequency range covered by the pods.

Besides testing the operation of the pods, Project Vampyrus experimented with techniques, trying to determine the formation that allowed the densest possible noise barrage without impacting the tactical effectiveness of the strike force. At first, the leader flew at altitudes between 6,000 and 16,000 feet, with others echeloned downward, each pilot remaining 1,000 feet below and 1,500 feet behind the aircraft ahead. These formations proved awkward, however, and the pilots complained that they found it hard to locate navigational checkpoints and targets because they had to look up so frequently to maintain station. To avoid this distraction, they decided to form an echelon extending upward from the leader. Vampyrus demonstrated that pod-carrying Thunderchiefs, flying in loose formation at medium altitude, could successfully jam radar-controlled weapons.

With losses mounting, 7th Air Force required that every F-105 going over North Vietnam carry ECM pods. By mid-October 1966, Takhli received their first batch of QRC-160A (later ALQ-71) pods[112] and started using them almost immediately. Korat received their first pods by the end of October. Reports from pilots on these early missions included "... orbiting over Kep at 18,000 feet with the flak bursting 10,000 feet below and SAMs going off ballistically but not guiding anywhere near the pod carriers."[113] Initially most F-105Ds were launched with an ALQ-71 pod on each outer wing station, multiple pods being necessary because of the limited frequency coverage afforded by each pod (different pods covered slightly different frequencies and polarities). The pods were marked ODD and EVEN (denoting the last digit of their serial numbers), and the rules were that the strike force had to carry an equal number of each type. One of the pods contained a modulator reinforcement tuned for the SA-2 elevation radar, while the other was set to counter the SA-2 azimuth radar. But when the MiG threat began to grow, the F-105 crews began using one of the stations to carry a single AIM-9B Sidewinder. Eventually a set of dual carriers was developed that allowed two ECM pods to be carried under one wing and two AIM-9s under the other.[114]

Colonel Robin Olds, commander of the Thailand-based 8th TFW, declared that "the most significant development in the air war over North Vietnam during my tour was the introduction of the ... ECM pod and with it a return to mass formation tactics reminiscent of fighter operations in World War II and Korea." Brigadier General William S. Chairsell, commander of the 388th TFW, made a similar report. "Prior to the ECM pod," he wrote, "our aircraft were required to ingress and egress to and from the target using terrain masking for protection and employ the 'pop-up' maneuver over the target," tactics that made them "extremely vulnerable to SA-2 firings and AAA at the peak of their pop-up." Using the ECM pods, General Chairsell continued, "our aircraft could now roll into the target from medium altitude—12,000 to 15,000 feet," a change that reduced losses and improved bombing accuracy.

Despite the impact on the loss rate, the ECM pods did not confer absolute immunity. During level flight, two pods protected an aircraft to within 8 to 10 nautical miles of a SAM site. Burn-through began to occur at closer distances as the Fan Song overpowered the jamming signal, enabling the controller to locate a target for his missiles, although this ceased to be a problem at very close ranges since the SA-2 had a minimum range of about 5 miles. Also, because of the pod's antenna pattern, jamming coverage decreased markedly during maneuvers, especially in steeply banked turns when the jamming might not blanket every radar-controlled weapon that could track the aircraft.

The ALQ-71 had a few design failings that became evident in early 1967. One was the location of the controls in the F-105D, which were installed slightly behind the pilot on the right console where it was difficult to see the pod failure indicator. The pod therefore might fail without the pilot realizing it. Maintenance crews in Thailand resolved the problem by moving the controls. At the same time, they corrected another failing, a matter of mutual interference, by rewiring the controls. Before this modification, the same jamming signal that disrupted enemy radar also prevented the pilot from using his RHAW equipment. The change in wiring enabled the pilot to interrupt the jamming signal for a few seconds.

Iron Hand teams had to exercise caution when using both jamming pods and antiradiation missiles because the ALQ-71 reduced the accuracy of the Shrike and created clutter on the RHAW scopes in Wild Weasel hunter aircraft. Iron Hand crews therefore tried to avoid turning on the pods until after receiving a SAM warning and launching an antiradiation missile.

In November 1967 SAMs scored eight kills within four days despite the ECM pods. An investigation team flew to Southeast Asia and discovered that some Fan Song radars were transmitting on a slightly lower frequency, thus escaping the jamming barrage. Also, aircraft tended to bunch up in order to obtain a more compact bomb pattern during missions controlled by Skyspot ground-based radars, and in doing so they simplified the task of tracking and aiming at the jamming source. Instructions immediately went out to open up formations and to adjust the center frequencies of the jammers directed against Fan Song.

Tracking the jamming source (also called passive tracking) enabled the North Vietnamese to diminish the effect of self-protection pods. Instead of relying on the radar return, which the jamming eliminated, the North Vietnamese launched missiles at the center of the jamming pattern. To determine range, the North Vietnamese used triangulation from widely separated radar sites linked by radio or telephone. An expert Fan Song operator would not energize the guidance signal until the Guideline had left the launcher. Although there was some sacrifice of accuracy, these tactics reduced the impact of the ALQ-71, decreased SAM warning time, and minimized exposure to antiradiation missiles.

The ALQ-87 (originally called QRC-160-8) began to arrive in Southeast Asia during late 1967. At first, a ram-air turbine supplied current for the four magnetron jammers housed in the ALQ-87, but this power source proved too fragile to withstand high-speed maneuvering. The pod was later powered directly from the aircraft's electrical system. Besides laying down a continuous jamming barrage, the ALQ-87 had a sweep modulator that could introduce random bursts of reinforcing noise in a so-called pulse power option. The pod could simultaneously perform any two of three functions: denying range and azimuth data to Fire Can; depriving Fan Song of range, altitude, and azimuth; and jamming the position beacon installed in the sustainer section of the Guideline missile.

Beacon jamming, also called downlink jamming, interfered with the signal that allowed the SAM controller to correct the missile's trajectory. Both the ALQ-71 and ALQ-87 enjoyed impressive success with this technique. For example, from the inception of beacon jamming in December 1967 until 1 April 1968, when bombing north of the 19th parallel was halted, SAM batteries launched 495 missiles at Iron Hand Thunderchiefs but only downed three aircraft, and two of these had been jamming the tracking beam instead of the downlink signal.

The next ECM pod to arrive in Southeast Asia was the Westinghouse QRC-335/ALQ-101, which began to arrive by July 1968. This pod could perform either deception or noise-barrage jamming, and was specifically intended for activities such as Iron Hand, where formation flying was not feasible. The ALQ-101 contained a self-destruct mechanism that destroyed the logic circuits when the pod was subjected to the excessive g-loads encountered on an impact. However, due to an engineering defect, several pods self-destructed while on the ground. An engineering change eventually eliminated this tendency. It appears that Thunderchief usage of the ALQ-101 was limited to F-105F aircraft at Korat for a short period of time in late 1968 when 14–22 pods were available for the 388th TFW. The ALQ-101 was later used extensively by F-4s.[115]

Although the flight manual, dated 8 March 1972, lists the ALQ-71(V)-2 and -3; ALQ-101(V)-3, -4, and -6; and ALQ-119(V)-1 as certified for use on the F-105D/F/G, it is unclear how many ALQ-101 or ALQ-119 pods were actually used by the F-105 squadrons. And even though the ALQ-87 was not listed at all, it is certain that the pod was used operationally by the F-105 in Southeast Asia.

Commando Nail

During January 1967, air attacks by the 355th and 388th TFWs against North Vietnam were seriously impeded by the thick cloud cover resulting from the annual monsoons. When General John D. Ryan became Commander-in-Chief of Pacific Air Forces on 1 February 1967, he immediately recognized the problem and decided that what was needed was the development of an all-weather and night radar bombing capability. In response to Ryan's request, a series of bombing tests were conducted at Nellis AFB under Operation Bullseye and evaluated by the Combat Target task force. These tests, which resulted in a report issued in Octo-

An F-105F-1-RE (63-8353) assigned to the 44th TFS at Korat in early 1968. Officially known as Operation Northscope or Commando Nail, these aircraft and their pilots were generally called Ryan's Raiders. Instead of the normal light gray undersides, the Raiders used a light green and tan to better hide the aircraft against the night sky. Under the nose is the APR-25 antenna fairing and the strike camera. Three of the four triangular az-el antennas can be seen spaced just behind the radome, while two stub antennas for the ER-142 are on the bottom of the fuselage just ahead of the nose gear. *(U.S. Air Force)*

ber 1967, obtained actual data on the bombing accuracies of the F-105D, F-105F, F-4D, and F-111A under a variety of conditions. These tests would form the basis for requirements that eventually resulted in greatly improved all-weather bombing systems, but they did little for the immediate need felt by General Ryan.

At Ryan's urging, PACAF initiated a special training program called Operation Northscope on 4 March 1967. The 41st Air Division at Yokota AB, Japan, was selected to develop a training program for 25 F-105 pilots to qualify them to fly all-weather bombing missions over North Vietnam in F-105Fs. The 41st was uniquely qualified for the task since they had experienced F-105 instructor pilots, and perhaps more importantly, maintenance personnel that could deal with the intricacies of the ASG-19 and its R-14A radar. Four F-105Fs that were just completing Wild Weasel III modifications were selected for use by the new program. But although their ECM and armament systems had been upgraded, their R-14As had not been thoroughly exercised in years.

From available documentation, it is unclear whether the four aircraft were modified concurrently with the first training course, or subsequent to it. It should be noted that the changes were developed and implemented locally, with little assistance from normal Air Force engineering and test organizations in the U.S. The 441st A&E Squadron in Yokota performed most of the modifications, with the remainder being accomplished by the 388th

TFW. Regardless, officially the modification was called T.O. 1F-105F-536, variously known as the "blind bombing mod" or the "all-weather bombing system."[116] Within the 388th TFW, the aircraft and crews were referred to as "Ryan's Raiders."

This modification made numerous small changes to the F-105F. The control stick, nuclear consent panel, clearance plane indicator follower, radar scope, ADI, and armament panel were removed from the aft cockpit. In their place a larger 5-inch radar scope, camera control panel, pedestal control panel, radar altimeter, and azimuth steering trimmer were added. The new radar scope, in addition to the improved resolution allowed by its slightly increased size, permitted a greater persistence (longer duration) that allowed the rear seat pilot more time to decipher the data. A moving radar cursor was added to minimize the tracking and freezing errors normally experienced.[117] The R-14A was modified to provide a 40° sector sweep in addition to the standard 90° sweep, and a new 20-mile radar range scale replaced the 13-mile scale in ground-mapping mode.[118] The narrow sector sweep reduced the period between successive radar sweeps, resulting in less target movement between sweeps. A new scope camera mounted over the radar scope in the front cockpit interfered with full-forward movement of the control stick, requiring the pilot to physically bend the camera mount out of the way when full control was needed.

The rear cockpit was equipped with an ER-142 panoramic receiver (added for the Weasel mission) located in the space normally used by the tape altimeter and airspeed indicator. The small backup altimeter and airspeed indicator used in place of the more accurate tape instruments presented somewhat of a problem since they were notoriously inaccurate. The pilot's weapons release switch was wired in parallel to the rear cockpit to allow the rear-seat pilot to control the bomb release. However, most of the changes involved fine-tuning already existing systems. The only external clue to the aircraft's new role was a special paint scheme. Instead of the normal light gray undersides, the Raiders used a light green and tan to better hide the aircraft against the night sky.[119]

An important change in concept was made for the original Ryan's Raider crews. Normally, the Wild Weasel F-105Fs carried an electronic warfare officer (EWO) in the back seat. For the most part these officers had been EWOs in SAC bombers before transferring to the Weasel program and had little, if any, piloting or navigator experience. Due to the demanding missions being planned, a decision was made that both cockpits of the Northscope F-105Fs would contain rated pilots, despite the fact that the flight controls were removed from the rear cockpit. This greatly speeded up the initial training since all F-105D pilots had received (and probably forgotten) low-level bombing training when they initially were assigned to the Thunderchief. The first 10 pilots selected for front-seat duties were instructor pilots with the 41st AD. Eight replacement pilots just arriving from the U.S. to begin tours with the 355th or 388th were diverted to serve as rear-seat pilots.

The training program hurriedly developed by the 41st AD was designed to give the rear-seat pilot maximum exposure to radar bombing techniques. After a 10-hour refresher course on the R-14A radar, the toss bomb computer, and radar interpretation techniques, the crews were given 12 flights within a short 20 hours of flying time. Each sortie was planned as a radar navigation exercise with simulated bomb runs along the route at medium altitudes between 10,000 and 15,000 feet. The simulated targets were industrial complexes and other prominent cultural radar targets that had been used often by Yokota pilots training for nuclear delivery. The training program was endorsed by PACAF with one significant exception—the flights should be conducted at altitudes below 2,000 feet.[120]

The low-level aspect of the missions was reportedly due to personal intervention by General Ryan. However, it was in keeping with the general belief that to penetrate North Vietnamese air defenses, aircraft needed to fly below the effective radar coverage. The altitude thought to be most effective was something below 1,000 feet, although this altitude probably exceeded the reasonable expectation from the R-14A radar's terrain-avoidance capability.

In any event, the initial reliability of the R-14A was so poor that little actual low-altitude training was accomplished. In fact, some training was completed over ranges in Korea when the weather in Japan prevented VFR flights. Nevertheless, the first four crews were certified in April and arrived at Korat on 24 April 1967 to form a provisional subunit commanded by Lieutenant Colonel Fred A. Treyz within the 13th TFS, usually known as the 13th Raiders.[121] Two days later, the first Ryan's Raiders mission struck targets on the Ron ferry complex and Yen Bai rail yards as part of Operation Northscope.

Normal weapons loading was standard for an F-105 in Southeast Asia—six M117 750-pound general-purpose bombs on the centerline MER, two 450-gallon drop tanks under the wings, and a pair of ECM pods on the outer wing stations. One of the M117s was usually configured with a 24-hour chemical delay fuze to provide a "surprise" the following night. Missions into high MiG threat areas would call for at least one AIM-9B Sidewinder missile on the right outer wing pylon, replacing one ECM pod. Although no MiGs were downed by Raider crews, two pilots were credited with searchlight "kills" using the heat-seeking Sidewinder missiles.

As with other F-105 operations in Southeast Asia, Ryan's Raiders frequently overloaded the airframe. With a full load, a Raider F-105F was at maximum gross weight before the fuselage fuel tanks were full. In daylight operations, the crews would simply take off and immediately rendezvous with a tanker to get a full fuel load. But night refueling was hazardous at best, and a real nightmare in bad weather, especially if the tanker was the probe-and-drogue type. Consequently, most Raider crews began to simply disregard the flight manual and launch at over maximum gross weight. They still had to refuel over northern Laos, but they only had to do it once.

A typical Raider mission started off with a radar calibration during the flight to the waiting tanker over northern Laos. If the radar and Doppler checked out, the crew would usually have to use the air-to-air mode of the R-14A just to find the tanker in the night skies over Laos. Following the refueling, the Raider pilot would proceed straight for the North Vietnamese border at about 10,000 feet, being tracked the entire time by North Vietnamese radars. Crossing the border, the Raider would begin a slow descent to about 450 knots at 500–1,000 feet. The pilot would change headings every five minutes en route to the target, always staying below ridge lines.

As the Raider approached the target, the pilot would accelerate to 500 knots and climb to 1,000 feet for bomb release. Following bomb release, the Raider pilot would select afterburner and egress the area at maximum speed, still remaining at low level. Once the back-seater had used the ER-142 to verify the aircraft was not being targeted by AAA or SAMs, the pilot would initiate a climb to a more fuel-efficient altitude. During this phase, an EB-66 from Takhli would orbit just outside North Vietnam to provide threat warnings of SAMs, AAA, and MiGs to the Raider at a much greater range than onboard systems could provide.

Four additional crews arrived on 8 May, and the last four dual-pilot crews arrived at Korat on 22 May. Together these 12 crews flew 98 missions into Route Pack 5 and 6A during the next 80 days. Targets included the Thai Nguyen steel mill, Kep airfield, Yen Bai rail yard, and the Bac Kan trans-shipment point. The missions came at a high cost. On 12 May 1967 an F-105F (63-8269) was lost when the aircraft simply disappeared into a poorly charted area of North Vietnam. The assumption, never conclusively proven, was that the crew exceeded the limits of the R-14A's terrain-avoidance capability and flew into the ground during the low-altitude penetration. Three days later a second F-105F (63-4429) was lost to heavy AAA fire after releasing its bombs over the Kep rail yards.[122]

Nevertheless, based on the initial success of the Raider missions, plans were made to expand the program. Eight additional aircraft were modified, although it appears that these were not Weasels. The modified aircraft were generally called 2098-aircraft, referring to the modification number. Additional crews were selected for training, but this time the crew composition changed. Initially it was thought that using a bombardier/navigator with more experience would enhance the mission. Plans were made to use radar-navigators from SAC B-58s who were very experienced in bomb delivery techniques and radar interpretation. However, SAC was reluctant to release any of the RNs.[123]

As a fallback position, since the seven surviving original aircraft were all configured as Wild Weasels, Lieutenant General Joseph H. Moore, Vice Commander of PACAF, decided that the next five crews should be qualified Weasel crews. This would allow the crew to perform normal Weasel operations in good weather, or Raider missions in bad weather. It was thought to be a sound idea for the effective use of crew personnel, and was a matter of economics in a war that was costing more than the Air Force could afford.[124]

There was, however, a flaw in the plan. The Weasel crews consisted of a qualified F-105 pilot and an EWO. The EWOs were assumed to be proficient radar operators since each was a rated[125] navigator. But in reality, none of the first five EWOs selected had ever flown as qualified navigators in any type of aircraft. Each had logged 1,340–2,800 hours as an EWO in SAC bombers but had never used radar for navigation, much less for bombing. The R-14A radar required minute refinements in antenna tilt and video gain to obtain the optimum returns necessary for accurate bombing, and this proved to be a challenge for the new crews. It would also prove a challenge for even experienced crews during low-level flights.

Again, the 41st AD at Yokota would provide the training for the five new crews. During the initial training flights it appeared that the EWOs were much slower to master the operation of the R-14A radar. This could easily be explained because none of them had ever operated an R-14A before, whereas the previous crews made up of pilots had been trained on the equipment when they qualified in the F-105. Surprisingly, after the initial learning curve, the EWOs soon surpassed the pilots in navigation and bombing accuracy.

The Weasel crews were all volunteers. Initially there was some resentment from the five crews selected for Northscope training because they believed this would prevent them from flying Weasel missions. However, once the crews were fully briefed on the nature and importance of Northscope, feelings began to change. Eventually, four of the five crews agreed to undertake Northscope missions, but the fifth crew elected to fly only Weasel missions. Training was completed on 17 July 1967 and the crews departed for Korat.[126]

The timing was excellent. Within a month of the four new Raider crews arriving, all of the original dual-pilot crews had either returned to day-strike duties or completed their combat tours and returned to the United States. A local plan was devised to rotate the four new crews between Raider and Weasel missions every 15 days to allow each crew to stay proficient in both missions. In reality, it allowed each crew to stay adequately accustomed to both missions, but probably denied them the opportunity to become truly proficient at either. The first missions were flown by the new crews as part of Operation Commando Nail on 15 July 1967 against targets in the relatively low-threat areas of Route Pack 1 since this was the first time either of the crews had flown night missions using the R-14A. The missions also represented another first—it was the first time an EWO had dropped a bomb in combat. All future Raider missions would be known as Commando Nail.

In July 1967, another mission was tasked to the Raiders by the 7th Air Force. In the past, bad weather in the form of a low-flying stratus deck in North Vietnam had curtailed air operations during the months of September through December. Major Kuster was ordered to develop tactics and procedures for high-altitude pathfinder operations. Known as Commando Nail Papa, combat evaluation missions began in late July. Initially the effort was restricted to periodic missions against the Ron and Quang Khe ferry complexes in Route Pack 1, flown when the Raider aircraft could be spared from northern targeting. During these flights two Raider aircraft would fly close formation, with the wingman releasing his bombs visually as soon as he saw the leader's bombs drop away. The EWO in the second aircraft, in addition to double-checking the release data, would film the impact with a handheld movie camera.[127]

During September 1967 the Raiders were assigned to attack the Somtra railroad siding located 20 miles northwest of Yen Bai along the Red River. These missions were flown as single and two-ship night attacks, and also four-ship day raids. A total of 29 bombs found the target. In reality these missions were a continuing evaluation of the accuracy of night versus day missions. The target had been selected because it had not previously been attacked (therefore having no previous craters), was militarily worthwhile, and was in a low-threat area. At the same time, F-4Ds were being evaluated under similar circumstances elsewhere. In the end, the evaluation showed that the F-105Fs achieved a circular error probability (CEP) of 2,910 feet, while the F-4D achieved 3,075 feet. The most accurate impact was reported as 800 feet to the left of the target. This left a lot to be desired since the mean effective radius of a 750-pound bomb is only 20–25 feet.[128]

During the evaluation, on 2 August 1967, one 16-ship mission led by Major Francis P. Walsh was flown against the army barracks near Phu Tho, North Vietnam. The purpose of the flight was to determine the feasibility of maneuvering a large number of aircraft while maintaining mutual ECM jamming support and a defensive posture against potential threats.[129] Four cells of four F-105s each were to bomb in cell with three seconds of separation between the flights. The leader of the three following four-ship cells had his bomb release timer set to three seconds. As soon as the first cell leader dropped his bombs, each of the following cells would trigger a release, which would be delayed three seconds by the TBC timer. The other three aircraft in each cell would drop their bombs as soon as they saw their leader drop his. In theory this would result in all the bombs impacting the same target area. Unfortunately, an otherwise perfect mission was marred by the fact that the lead aircraft's radar was not accurately calibrated. All of the bombs dropped in close proximity to each other, but a mile from the intended target. It validated the concept nevertheless, and all aircraft assigned to the Raider mission had their radars recalibrated before their next mission.[130]

The pathfinder missions would continue for the duration of the Commando Nail. Generally, a pair of Raider F-105Fs would lead a standard daylight formation of 16 aircraft on missions into Route Pack 5 and 6A, where the weather was not conducive to visual bombing rules. Although successful, the pathfinder missions left themselves open to SAM attacks, since the formation had to spread out to obtain an acceptable bomb pattern. But spreading the formation also spread the ECM coverage, and flying above the clouds did not allow visual acquisition of a SAM until it broke through the cloud deck. And by then it was too late for evasive action.

On 4 October 1967, low-level Commando Nail missions were halted when another F-105F (63-8346) was lost during a night attack on the Phu Tho rail yard near the Chinese border. This loss triggered a revaluation of the merits of the low-altitude missions and proved to be the end of the single-ship low-level penetrations. During the period from 26 April to 4 October 1967, the Raiders had flown 415 sorties in Route Packs 1, 5, and 6A. In any case, beginning in August many of the Raider aircraft had been diverted to Wild Weasel support of B-52 missions. In reality, the remaining dual-purpose aircraft reverted exclusively to Wild Weasel duties, although they remained available to Ryan's Raiders as replacements for aircraft being repaired.[131]

Following the 4 October loss, Raider missions were restricted to interdiction raids into Route Pack 1 in southern North Vietnam. Route Pack 1 had been used as a training area for the Raider crews since the air defense network in the area was not highly integrated and allowed some latitude in operations. Previous missions into Route Pack 1 had been used by the Raider crews to practice low-level flying techniques and to perfect radar bombing procedures prior to venturing north. These were still night missions and very hazardous, but only required in-flight refueling if they were along the eastern coastal areas. Following the termination of the northern missions, the Raiders generally flew two sorties per night into Route Pack 1 for the duration of the bombing campaign. Generally the targets were truck parks and supply points defended by relatively unsophisticated air defenses.

The more relaxed atmosphere for operations paid off in greater effectiveness in radar bombing. The crews could choose any angle of attack necessary to obtain the best radar picture since very little deception was needed to avoid defensive positions. This also allowed the Raider crews the first opportunity to assess their own effectiveness, since the mission over the north had allowed little opportunity to orbit the target area to observe the bomb impact. In October 1968 four more Raider crews, these trained stateside at McConnell AFB, arrived. Like the other recent crews, these consisted of a pilot and EWO, even though the dual-purpose aircraft had essentially been withdrawn from Raider missions.

Sometimes luck was with the Raiders. On 25 October a Raider attacked a known truck park in the Co Ta Roun mountains, 3 miles from the Laotian border. A 5,325-foot mountain peak 5 miles south of the area offered an excellent radar offset aiming point, and minimal air defenses allowed the Raider to attack from 10,000 feet with sufficient time to obtain a nearly perfect radar return. The aircraft went inverted after bomb release to allow the crew to observe the effects. A large orange flash, accompanied by black smoke billowing up 3,000 feet offered testimony, it was thought, to the accuracy of the bombing.

An RF-4C was sent over the target to record the damage, and copies of the photos made it to Korat two days later. The photos showed that the truck park was intact and untouched. However, a small hill 600 feet away was virtually leveled and showed indications of a large, intense petroleum fire. Obviously a fuel depot had been hit by a quirk of fate.[132]

In November 1968, the Ryan Raider mission ended, although interestingly, four additional crews arrived in early December. However, even with its limited systems and crews, the mission of Ryan's Raiders was deemed a success, and it led to future deployments of F-111A single-aircraft missions, as well as night interdiction missions flown by the F-4D-equipped 497th TFS "Night Owls." Other units flying F-4Es would also pick up some of the prior Commando Nail missions.

With their mission gone, a decision on the disposition of the 2098-modified F-105Fs needed to be made. Since most of the surviving aircraft were not Wild Weasels, they could not be assigned to that role. The training squadron at McConnell AFB needed additional F-105F aircraft, and it was decided that the aircraft would be returned to the United States for use as transition trainers. The 44th TFS (which was the successor squadron to the 13th TFS) did not necessarily agree with this decision. The 44th wanted to keep the aircraft and continue to use them on pathfinder missions, much like Commando Nail Papa. This was largely in response to maintenance problems being experienced by Combat Skyspot and the desire of the 388th to become independent of Skyspot. The 44th pointed out that the bombing accuracy had been much better under Commando Nail Papa (400-foot CEP, versus 1,800-feet normally), and that these aircraft greatly enhanced the Wing's combat effectiveness and capability.[133]

The 44th TFS made a point that 2098 was a Class V modification in which the rear cockpit control stick and part of the instruments had been removed, and therefore the aircraft could not be flown from the rear seat. In order for these aircraft to be used for training with instructor pilots aboard, funds would be required to return the aircraft to their original F-105F configuration. If the demodification was not accomplished, the 2098 aircraft could only effectively be used as a single seater. Nevertheless, five 2098-modified F-105Fs left Korat for McConnell AFB on 10 December 1968, destined to augment TAC's combat crew training school for strike pilots. Eventually funds were found to demodify the aircraft.[134]

When the war started to wind down after the late 1968 bombing halt, many of the Combat Martin F-105Fs were transferred to other squadrons still flying the F-105. This Combat Martin F-105F-1-RE (62-4432), assigned to the 561st TFS at McConnell AFB in 1970, was later modified to the F-105G configuration prior to deployment during Linebacker II. *(Joe Bruch via Larry Davis via the Mick Roth Collection)*

Combat Martin

During mid-1967 the Air Force conducted a comparative evaluation of the Navy's Sanders Associates AN/ALQ-55 communications jammer and a Hallicrafters VHF communications jammer developed under QRC-128, subsequently redesignated AN/ALQ-59. Both systems were installed in an F-105F for the evaluation, which may have been conducted under QRC-363. The purpose of both systems was to jam North Vietnamese VHF ground-control intercept (GCI) communications. The ALQ-55 was used by a variety of Navy aircraft, while the ALQ-59 was also used by the EB-66.[135]

The ALQ-59 was selected for installation on 10 F-105Fs under Combat Martin. The ejection seat and instrument panel in the rear cockpit were removed and replaced by the ALQ-59 VHF communications jamming system. Externally, a Combat Martin F-105F could be distinguished by the single large-blade antenna mounted on the fuselage spine just aft of the rear cockpit and the absence of the normal ejection seat in the rear cockpit. Some official sources from the Vietnam era list these aircraft as EF-105Fs, but it is unlikely this designation[136] was officially approved. Combat Martin aircraft were reportedly assigned to both the 355th and 388th TFW from December 1967 until the beginning of 1970.

Interestingly, it appears that Combat Martin F-105Fs never flew in their intended role since they were never authorized to employ the ALQ-59 over North Vietnam. Reports indicate that the only missions flown by the aircraft were standard bombing missions, flown by a single pilot in the front seat and without the ALQ-59 equipment in the back seat (although the antenna was retained). This was not a failing in the equipment, but a simple matter of making the best use of a limited asset (the F-105) since the comm jamming missions were

A Combat Martin F-105F-1-RE (62-4444) from the 357th TFS shows the large blade antenna on top of the fuselage that was installed for the ALQ-59 communications jammer. Note the empty back cockpit—not even an ejection seat. The Combat Martin aircraft were apparently never employed in their intended roles since authorization for the missions was never granted from higher headquarters. However, the aircraft were flown on normal bombing missions, after the major portions of the ALQ-59 jamming systems were removed. The aircraft also carries an ALQ-87 ECM pod and an empty MER. *(U.S. Air Force/DVIC)*

never approved by higher commands. However, because of the high-value status[137] of the modified aircraft, the F-105Fs were under restrictions regarding bomb release altitudes and strafing runs. When the 355th stood down in November 1970, the Combat Martin aircraft were modified to Wild Weasel configuration and brought up to the F-105G standard.

Combat Skyspot

The navigation systems of the Vietnam era were proving to be less accurate than desired, especially at higher altitudes. To guide tactical aircraft to their targets, the Air Force developed the AN/MSQ-77 Radar Ground Bombing Central and a technique called ground-directed bombing (GDB). The radar was a derivative of the bomb-scoring system used by SAC. The X-band MSQ-77 had a range of approximately 200 miles and was equipped with a vacuum-tube-type computer and a plotting board, which could draw a map of where an aircraft was flying. These maps could precisely determine where an aircraft was in relation to a chosen target. The computer continuously calculated the altitude, airspeed, wind drift correction, and ground elevation changes, using the ballistics of the bombs that were being carried by the aircraft. The plotting board operators provided steering corrections by voice to the attacking aircraft, and a bomb release tone was transmitted at the appropriate time.[138]

The SST-181X transponder was developed to improve tracking and extend the effective range of the radar site and was installed on some F-105D/Fs and F-4s, as well as various B-52s. The F-105 modification was carried out under TCTO 1F-105D-692 and 1F-105F-546 (Installation of SST-181X Transponder/Antenna System), and it appears that 250 Ds and 45 Fs were modified. The SST-181X control panel was located on the top of the main instrument panel and incorporated the transponder ON/OFF switch and an indicator light. The transponder was a solid-state system that increased the range-tracking capabilities of the MSQ-77 radars. The ground-based radars interrogated the SST-181X radar beacon, which

Another Combat Martin F-105F-1-RE (63-8291) shows the large UHF antenna and empty back cockpit. Reportedly the ground crews began calling the ALQ-59 "Colonel Computer" after somebody painted the name on the rear canopy rail of one aircraft. Since their communications jamming mission was denied them, the aircraft were generally employed as single-seat strike aircraft. *(via Larry Davis via the Mick Roth Collection)*

received and retransmitted a reply of much greater signal strength (same frequency). Hence, the radar site actually received a considerably stronger signal than a radar echo, improving target acquisition capabilities at long ranges. The Combat Skyspot missions were flown under the name Commando Club.[139]

A small number of MSQ-77 precision radar sites were established in South Vietnam and Laos. Perhaps the best known of these was Lima Site 85, established atop a steep 5,500-foot ridge called Phou Pha Thi. This mountain was located 15 miles from the North Vietnamese/Laotian border and 28 miles west of Sam Neua, Laos. A TACAN facility had already been established atop Phou Pha Thi and was operational in August 1966. The AN/TSQ-81 equipment installed at Site 85 differed from the MSQ-77 primarily in that it was packaged in a manner that made it easier to transport via helicopter, given the logistical considerations of the mountain site. The TSQ-81 was functionally identical to the MSQ-77.

The TSQ-81 facility at Site 85 was established to help enhance USAF all-weather strike capabilities against the northern route packages in North Vietnam and targets in Northeastern Laos. Since the weather over North Vietnam generally turned unfavorable for air operations in mid-October and did not begin to improve until April, it was considered imperative that the site be operational before the weather deteriorated. The installation crews came close, and the site became operational at the end of October 1967. Excluding Route Package I, the following data indicate the use of the Site 85 TSQ-81 facility in directing actual strikes against North Vietnam. This data includes all aircraft, not specifically F-105s.

The left side of the Desert Fox F-105D-31-RE (62-4299) photographed at China Lake, California. All markings, except for the yellow fin stripe, were painted in black. Barely visible is a desert fox drawing on the fuselage under the air intake. Note the 419th TFW markings on the nose wheel door. *(Mick Roth)*

	Nov.	Dec.	Jan.	Feb.	Mar. 1–10	Total
Total Missions	153	94	125	49	6	427
Missions under Commando Club	20	20	29	27	3	99
Percentage under Commando Club	13.0	21.3	23.2	55.1	50.0	23.2

Apparently the North Vietnamese were well aware of the techniques being used. Although great care was taken to protect the location of Site 85, on 13 January 1968 the North Vietnamese used two AN-2 Colts to attack the radar installation. The damage from this first attack was relatively minor, and the site continued to operate both the TSQ-81 and TACAN equipment, even though it was known the Vietnamese were preparing to launch a ground attack on the site. The Air Force and CIA directed numerous air strikes of F-4s, F-105s, and A-1s from Thailand and Vietnam, many using the radar at Site 85, against the massed columns of enemy appearing to encircle the site. The strikes were increased, even using Air Commando A-26 Invaders to attack at night, in an attempt to turn the twin advances on Routes 6 and 19. The ground attack finally came on 11 March 1968, destroying Site 85 and killing 10 of the 16 Americans still based there. Combat Skyspot continued until the end of the war, using radars at other sites.[140]

The End

Over 20,000 combat missions were flown by Thunderchiefs in Vietnam. At one point in 1965–1968, it was calculated that an F-105 pilot stood only a 75 percent chance of surviving 100 missions over North Vietnam. Although the total number of losses was rather high, the actual loss rate was not that bad considering the total number of missions that were flown. In late-1967, the Air Force admitted it had lost 330 F-105s over North Vietnam, resulting in

The end of the line. An F-105D-25-RE (61-0183) from the 563rd TFS is prepared for storage in the Military Aircraft Storage and Disposition Center (MASDC) at Davis-Monthan AFB, Arizona, during September 1971. All openings on the aircraft were sealed to prevent dust and moisture from entering. Although many aircraft are eventually returned to active duty after being stored in MASDC, the F-105s would not be among them. *(Terry Panopalis Collection)*

a loss rate of 1.6 aircraft per 1,000 sorties. By August of 1968, so many F-105s had been lost that the squadron strength of the 355th and 388th TFWs was reduced from the normal 24 aircraft each to just 18. In less than 10 years, the total strength of 833 production F-105s had been cut by battle losses and general attrition to 427, and 60 of those were B-models that were not considered combat capable.

Most F-105Ds were withdrawn from Southeast Asia by October 1970 as F-4s took over the fighter-bomber role, but the Wild Weasels would stay in Vietnam until the end. Besides, there were not that many F-105Ds left, and the type earned the distinction of being the first aircraft to be withdrawn from active Air Force service due to attrition. Following their withdrawal from Southeast Asia, the few F-105Ds to survive combat served with active duty Air Force units for a couple more years, primarily with the 23rd TFW at McConnell AFB and the 18th TFW at Kadena AB on Okinawa. The official phase-out began in November 1970 when two ANG units were notified they would be receiving the remaining F-105Ds. Aircraft began reaching the 184th Tactical Fighter Training Group (TFTG) at McConnell AFB, Kansas, and the 192nd Tactical Fighter Group (TFG) at Byrd Field, Virginia, in January 1971. Conversion of the 113th TFG at Andrews AFB followed later in the year.

In a reversal of then-recent policy that had limited the Air Force Reserve to airlift duties, the 507th TFG at Tinker AFB, Oklahoma, began to receive F-105Ds in June 1972. The 301st TFW at Carswell AFB, Texas, followed and began receiving F-105Ds in August 1972. By June 1973, only six F-105Ds were left on active service—two for special testing at Eglin, and four used for training Wild Weasel pilots. The surviving F-105Ds were delivered for storage at Davis-Monthan AFB, and although many ended up being turned over to museums, none are still flying today. Interestingly, in a message posted to various forums on the Internet during the summer of 1999, a South African group is trying to purchase an F-105 and J75 to restore it to flyable condition for the air show circuit.

Notes

91. During 1999 it was publicly admitted by various officials from the time that this second attack probably never happened. A combination of miscommunications and general confusion led to the initial report of the attack, and political necessity propagated the news even after it had been officially questioned.
92. Davis, Larry and Menard, David, *Republic F-105 Thunderchief,* Volume 18 in the WarBird Tech Series, Specialty Press, North Branch, MN.
93. These losses only include those directly attributed to combat. There were also losses to accidents that raised these numbers considerably.
94. *Pacific Air Forces: Fifty Years In Defense of the Nation,* p 80.
95. Davis, Larry and Menard, David, *Republic F-105 Thunderchief,* Volume 18 in the WarBird Tech Series, Specialty Press, North Branch, MN.
96. Anderson, Jr., Clarence E, Colonel, *End of Tour Report, 355th Tactical Fighter Wing,* 6 October 1970, p 2–6.
97. *Ibid.,* p 5.
98. *Ibid.,* p 9.
99. *Ibid.,* attachments.
100. Davis, Larry and Menard, David, *Republic F-105 Thunderchief,* Volume 18 in the WarBird Tech Series, Specialty Press, North Branch, MN.
101. *Ibid.*
102. Anderson, Jr., Clarence E, Colonel, *End of Tour Report, 355th Tactical Fighter Wing,* 6 October 1970, p 9.
103. Davis, Larry and Menard, David, *Republic F-105 Thunderchief,* Volume 18 in the WarBird Tech Series, Specialty Press, North Branch, MN.
104. SEA Operational Requirement Status Report, 25 February 1966.
105. 1F-105D-1 Flight Manual, 30 June 1969, changed 9 September 1970; *100 Missions North,* Kenneth H. Bell, Brassey's (US), 1993, p 19.
106. *355th TFW Brochure, the Vector in a SAM Environment,* 7 May 1966, p 7.
107. *Ibid.,* p 7.
108. Davis, Larry and Menard, David, *Republic F-105 Thunderchief,* Volume 18 in the WarBird Tech Series, Specialty Press, North Branch, MN.
109. *Tactics and Techniques of Electronic Warfare: Electronic Countermeasures in the Air War Against North Vietnam, 1965–1973,* Bernard C. Nalty, Office of Air Force History, 16 August 1977.
110. 388th TFW Combat Tactics, 10 September 1967, pp 3–5.
111. Interview with Ed Rasimus via email, 23 August 1999.
112. The only operational difference between the QRC-160A-1 and ALQ-71 was that the QRC pod only required 1 minute to warm up, versus 3–5 minutes for the ALQ-71.
113. Interview with Ed Rasimus via email, 23 August 1999.
114. Davis, Larry and Menard, David, *Republic F-105 Thunderchief,* Volume 18 in the WarBird Tech Series, Specialty Press, North Branch, MN.
115. 44th TFS History, 1 November–31 December 1968.
116. Michael, Major, Albert L., *Corona Harvest, Ryan's Raiders: A Special Report,* Aerospace Studies Institute, Air University, Maxwell AFB, Alabama, January 1970.
117. The normal R-14A used a fixed cursor that depended upon the B/N depressing the "freeze" button right as the target passed under the cursor. With the moving cursor, the B/N could move the cursor on top of the target independently of aircraft movement, greatly improving accuracy.
118. The indicator light on the control panel still said 13 miles.
119. Operational supplement 1F-105D-1S-91 to the F-105D/F Flight Manual, 8 March 1968.
120. Michael, Major, Albert L., *Corona Harvest, Ryan's Raiders: A Special Report,* Aerospace Studies Institute, Air University, Maxwell AFB, Alabama, January 1970, p 3.
121. This provisional unit would be absorbed into the 13th TFS on 1 June 1967, by which time it was commanded by Major Ralph L. Kuster, Jr.
122. Davis, Larry and Menard, David, *Republic F-105 Thunderchief,* Volume 18 in the WarBird Tech Series, Specialty Press, North Branch, MN.
123. Michael, Major, Albert L., *Corona Harvest, Ryan's Raiders: A Special Report,* Aerospace Studies Institute, Air University, Maxwell AFB, Alabama, January 1970, p 8.
124. *Ibid.,* p 6.
125. Flight crewmen had to be rated. The only choices were pilot and navigator, so SAC simply called the EWOs navigators.

126. Michael, Major, Albert L., *Corona Harvest, Ryan's Raiders: A Special Report,* Aerospace Studies Institute, Air University, Maxwell AFB, Alabama, January 1970, p 7.
127. 13th TFS Historical Data Record, 1 April 1967 through 31 July 1967, p 4.
128. Michael, Major, Albert L., *Corona Harvest, Ryan's Raiders: A Special Report,* Aerospace Studies Institute, Air University, Maxwell AFB, Alabama, January 1970, p 26.
129. 13th TFS Historical Data Record, 1 August 1967 through 31 August 1967, p 2.
130. Michael, Major, Albert L., *Corona Harvest, Ryan's Raiders: A Special Report,* Aerospace Studies Institute, Air University, Maxwell AFB, Alabama, January 1970, p 29.
131. *Tactics and Techniques of Electronic Warfare: Electronic Countermeasures in the Air War Against North Vietnam, 1965–1973,* Bernard C. Nalty, Office of Air Force History, 16 August 1977, p 50.
132. Michael, Major, Albert L., *Corona Harvest, Ryan's Raiders: A Special Report,* Aerospace Studies Institute, Air University, Maxwell AFB, Alabama, January 1970, pp 24–25.
133. Letter from Lt. Col. Sherrill, 44th TFS, to DCO, 388th TFW, December 1968.
134. 388th TFW History, October–December 1968.
135. The exact nature of QRC-363 could not be determined, and this is just one possibility.
136. Some sources also list the Wild Weasel III aircraft as EF-105Fs, but that application was specifically disapproved.
137. There were only ten aircraft, and the loss of any of them would have affected Combat Martin performance if the jamming missions had ever been approved.
138. Nellis AFB Web site.
139. Op Supplement 1F-105D-1S-45, 13 February 1967; F-4-34-1-1, 1976.
140. Project CHECO report, *The Fall of Site 85,* Captain Edward Vallentiny, 9 August 1968.

Wild Weasels—The F-105F/G

The Rolling Thunder bombing campaign began the first week in March 1965; scarcely a month later, on 5 April, a 100th SRW U-2 operating from Bien Hoa AB photographed an SA-2 site under construction 15 miles south-southeast of Hanoi. By early May, two sites had been completed in the Hanoi area and others were under construction there and in the vicinity of Haiphong. AAA sites were also being built adjacent to the new SAM sites for close-in defense against air attack.

The Joint Chiefs of Staff requested permission to strike the sites before they became operational, but Secretary of Defense Robert McNamara refused. The Johnson administration believed that the North Vietnamese would not fire their SAMs at U.S. aircraft unless the sites were attacked. That the North Vietnamese had acquired an advanced defensive weapon system to actually *use* was incomprehensible to McNamara's analysts. The prevailing opinion at the policy level seemed to be that by building SAM sites the North Vietnamese—with Soviet assistance—were only posturing. By exercising restraint and not attacking the sites, the U.S. would signal the North Vietnamese not to use them. This rosy assessment was encouraged by the administration's reluctance to provoke the Soviet Union by attacking facilities at which their advisors were certain to be present.

Daily surveillance of the sites by U-2s, Ryan Model 147B drones, and tactical reconnaissance aircraft continued, but cloud cover frequently limited high-altitude photographic coverage. Photo reconnaissance was supplemented by RB-66C ELINT missions in the Hanoi and Haiphong areas. By early July, four SAM sites had been completed around Hanoi and others were under construction. Finally, on 23 July, an RB-66C intercepted and recorded Fan Song tracking radar signals from one of the Hanoi-area sites.

On 24 July a MiGCAP flight of four 47th TFS/15th TFW F-4Cs supporting a strike on the Lang Chi munitions factory became the first aircraft in Southeast Asia to be fired upon by SAMs. One Phantom was downed and the other three were damaged by shrapnel. The following day a SAC reconnaissance drone was also downed by an SA-2.[141]

On 27 July 1965, U.S. forces flew the first Iron Hand (anti-SAM) air strikes of the war when 46 F-105Ds, escorted by 8 F-104Cs and 12 F-4Cs, struck the two sites that were thought to have fired on 24 and 25 July. In order to stay below the SA-2's assessed operational envelope, the F-105s approached the sites at low altitude and attacked at altitudes as low as 50 feet. Four F-105Ds were lost to automatic weapons fire over the targets and another, limping back to Thailand with battle damage, pitched out of control and collided with an escorting Thunderchief. Both went down. The attack had cost six F-105s (two from each wing), and only one of the pilots was rescued. Two days later, a 45th TRS RF-101C was also downed by antiaircraft fire from one of the sites.

Poststrike reconnaissance photographs revealed that the attacks had damaged neither site appreciably. Further, 2nd Air Division photo interpreters discovered that the missiles and Fan Song van at one site were dummies and the other site was unoccupied. Anticipating retaliatory air strikes, the North Vietnamese had evacuated the sites but left in place the heavy concentrations of automatic weapons surrounding them. The site baited with dummy equipment then became an effective flak trap.

This newly revealed mobility of the SA-2 and its Fan Song radar vans took U.S. analysts by surprise. Once an occupied site had fired at U.S. aircraft or been overflown by a reconnaissance aircraft, the North Vietnamese quickly moved the missiles and radar van to another site. In belated recognition of this mobility, the Air Force established quick-reaction F-105 strike flights at Korat and Takhli to respond once a SAM site revealed itself. However, before they could attack a site the alert force pilots needed more than just its general location; an RF-101 had to be dispatched to photograph the area in an attempt to pinpoint the site's exact position. The process of photographing the site, developing and examining the film, and delivering the prints to the alert force took up to eight hours. By then the North Vietnamese had moved the SAMs. The quick-reaction alert force concept proved, in actuality, to be anything but quick, and within a few weeks the practice was discontinued. In the future it would become North Vietnamese practice to construct large numbers of fixed SA-2 sites and rotate their SAM battalions among them. The North Vietnamese would also begin to camouflage active SAM sites.

The next victim of a SAM was downed on the night of 11–12 August 1965 when SAMs were launched against two Navy A-4Es on an armed reconnaissance mission about 50 miles south southwest of Hanoi. One A-4 was destroyed; the other returned to the USS *Midway* (CVA-41) with shrapnel damage to the starboard wing, tail, and fuselage. In response, Carrier Task Force 77 launched multiple Iron Hand missions on the night of 12 August and throughout the following day. No SAM sites were even located, let alone destroyed, but five aircraft were downed by AAA. Four were damaged by gunfire but returned to the *Midway* and *Coral Sea* (CVA-43). Vice Admiral Malcolm W. Cagle later wrote "it was truly a black Friday the 13th for TF-77."[142]

In August the Air Force tried a new method of locating SAM sites: using a Ryan 147 drone to stimulate the North Vietnamese air defense system, enabling ELINT aircraft to plot the location of any SAM site that responded. The first attempt, on 21 August, failed to draw a response. Ten days later a second attempt drew responses from three sites. Four F-105Ds armed with napalm canisters were then dispatched from Korat to attack one of the sites, but the ELINT fix was insufficiently precise to allow them to locate it. Low on fuel and unable to locate the camouflaged site, they finally attacked an alternate target, a wooden bridge, where heavy automatic weapons fire downed one F-105.[143]

The SA-2 represented a significant threat to U.S. air operations in Southeast Asia, and one that could not be countered with currently available equipment and tactics. On 12 August 1965 the Joint Chiefs of Staff formed a committee, code-named Prong Tong, to examine the use of SAMs in North Vietnam and determine the best means of countering them. The following day a special Air Force Anti-SAM Task Force, composed of representatives from the major air commands and headed by Brigadier General Kenneth C. Dempster, was also formed to find a solution or solutions to what was termed the "DRV [Democratic Republic of

Vietnam] SAM problem." The task force's findings included a number of recommendations, including the high-priority development of improved radar warning gear, self-protection ECM systems, and more effective weapons, but the key recommendation was that an independent, specialized, hunter-killer force equipped with existing countermeasures-receiving equipment be organized, trained, and deployed to Southeast Asia as quickly as possible to locate and attack SA-2 sites on a continuing basis. Because of the significant threat posed by antiaircraft guns surrounding most SA-2 sites, surprise was considered to be of paramount importance. SAM suppressors should not employ jamming, it was suggested, since it would alert site defenders that an attack was imminent. Instead, high-speed, low-altitude attacks by two-ship elements, with a maximum of two elements per site, were recommended. Once the specially equipped pathfinder aircraft had located and marked a SAM site with 2.75-inch rockets, accompanying F-105 Iron Hand aircraft would destroy it with cluster bombs, napalm, and 20-mm cannon fire.[144]

Once the hunter-killer concept had been developed, a search began for the best available equipment and platform for the mission. By the end of August North American Aviation (NAA) had been selected to integrate two Applied Technology, Inc. (ATI) systems, the Vector IV RHAW set and the IR-133 panoramic receiver, into the two-seat F-100F under the code name Wild Weasel.[145] In early October the Air Force approved the addition of a third system, the ATI WR-300 launch warning receiver, to the Wild Weasel avionics suite. Although the F-100F was not the best aircraft for the SAM suppression mission, it was more available for a test program than its F-105 two-seat counterpart, the F-105F, and had adequate performance to serve as a proof-of-concept test bed.

Wild Weasel II

One of the RHAW systems considered and rejected for the F-100F was Bendix-Pacific's pod-mounted APS-107 (XA-1), which was being installed in a test-bed F-105D (61-0164) at Republic's Farmingdale plant when the North Vietnamese SA-2s claimed their first victims in July. The system, which was intended for use on the F-105 fleet, was originally scheduled for an initial engineering flight test and evaluation in September 1965, but Air Force Headquarters changed the game plan in an effort to field the APS-107 as quickly as possible. The routine engineering evaluation became a combined engineering and operational test intended simultaneously to prove the system and develop tactics for its use.

When flights of the modified aircraft began on 5 August, they quickly revealed a major flaw in the APS-107 installation. Republic engineers had designed an installation that appeared to be both simple and effective in terms of antenna coverage: the system's eight antennas and their associated radio frequency detectors, video assemblies, and power dividers were installed in wingtip pods. At even moderate airspeeds, however, aerodynamic loads caused the pods to flutter. The severe vibration not only damaged the system's electronics but was also a significant safety-of-flight issue. After only a few flights, the Air Force suspended testing while Republic conducted stress and vibration analyses. Because of the urgency of the test, the Air Force elected to have Republic beef up the F-105's wings as a temporary measure, then resume testing with the aircraft limited to airspeeds of less than 300 knots. The required wing modifications delayed testing until late September, by which time Republic still had not come up with a permanent fix. When testing resumed, it quickly became apparent that while the APS-107 showed potential as a RHAW system, it was too immature for immediate service use. Fixes were required in the system itself, as well as in its installation in the F-105.[146]

During late September and early October, Air Force Headquarters consolidated the development and testing of applicable ECM equipment, weapons, and tactics under the title of Anti-SAM Tactics and Capability Program. On 11 October a message from the Air Force Chief of Staff brought the F-100F Wild Weasel project, Bendix DPN-61 homing system, and APS-107 under the Anti-SAM Tactics and Capability umbrella, along with 10 other projects.[147]

By mid-October the first two Wild Weasel F-100Fs, equipped with ATI's Vector RHAW system, were being flight tested at Eglin AFB with encouraging early results. Meanwhile, the F-105D/APS-107 (XA-1) project was dead in the water while Bendix-Pacific redesigned the pod-mounted system for internal installation. The modified system, designated APS-107B (X-1), would not be ready for testing until mid-to-late November. Such was the urgency to obtain RHAW equipment for the F-105D that the delay essentially removed the APS-107 as a candidate system. While the Melpar APR-23B had at one time been considered as a possible RHAW system for the F-105, it could only be installed in place of the aircraft's radar and fire control system, a trade the Air Force was not willing to make. Furthermore, Melpar was experiencing severe problems in producing a functional production version of the system for use in the F-4C. Consequently, on 15 October Air Force Chief of Staff General John P. McConnell dispatched a message to the Air Force Logistics Command, Sacramento Air Materiel Area (SMAMA), and TAC, which changed the direction of the F-105 RHAW program:

> **Radar homing and warning capability for tactical fighter aircraft is a critical requirement for SEA. Extraordinary measures are in order. The F-105, in its role as the Rolling Thunder air-to-ground workhorse, is experiencing an abnormal attrition rate. To reduce this loss rate, multiple simultaneous prototyping and testing is required to expedite selection of a RHAW system for the F-105 aircraft. The F-100 Vector test results have been very successful. If the internally mounted Vector antenna is compatible with the F-105 radome, it appears this would be the most expeditious and lowest cost alternative for the F-105. An expedited compatibility and operational capability determination is needed for the Applied Technology, Inc. Vector system in an F-105D. . . . The target date for flight of this system is ten calendar days after receipt of this message. This Headquarters desires that this be done in-house with existing resources. . . .**[148]

SMAMA, assisted by ATI, responded immediately and began working around the clock to transplant the F-100F's Vector installation into the F-105D. Not only did the SMAMA/ATI team succeed, but they managed to do it in only five days. The first Vector-equipped F-105D (62-4291) made its first flight at McClellan AFB on 20 October 1965, and five days later it was flown to Eglin AFB for testing.[149]

The original intent of the F-105D test was to evaluate the Vector IV as a substitute RHAW system for the F-105, but there was considerable interest in evaluating the capability of a single-seat aircraft to perform the Wild Weasel mission. Work had barely begun on the Vector F-105D when SMAMA was directed to equip another F-105D (61-0138) with an integrated system consisting of the Bendix DPN-61 homing receiver and the Maxson fin cap radar warning receiver. Again, the deadline was 10 days. The modification was completed on 27 October and the aircraft delivered to Eglin the following day.[150]

Such was the urgency to equip tactical aircraft in Southeast Asia with RHAW equipment that the Air Force didn't wait for the results of the F-105D test before ordering the Vector IV into production. Based on the F-100F Wild Weasel testing to date, the Air Force decided to procure 500 Vector IV sets and an equal number of the as-yet untested WR-300 launch warning receivers under the military designations APR-25 and APR-26, respectively. This occurred on 22 October, the day the Vector-equipped F-105D arrived at Eglin AFB.[151]

At about the same time, the Air Force redesignated the F-100F project Wild Weasel I and established Wild Weasel II as an F-105 umbrella project for the two modified F-105Ds and a still-to-be-modified F-105F (62-4421) which was slated to be equipped with the APS-107B (X-1) and ALQ-51—when they became available—and provisions for QRC-160-1 ECM pods.[152] Wild Weasel II had been in existence for only a few days when the Tactical Air Warfare Center (TAWC), in a 1 November message, asked Air Force Headquarters to approve a change in the project's scope:

The first F-105B-1-RE (54-0100) carried colorful markings for the first part of its career as a test aircraft. The wings, vertical stabilizer, and stabilators had red trim, and the rear of the cockpit fairing was red instead of the normal black. Note the inside of the landing gear door is also red. *(Craig Kaston Collection via the Mick Roth Collection)*

This is how the B-models spent most of their careers—in Southeast Asia camouflage. This F-105B-20-RE (57-5814) was assigned to the 466th TFS, Air Force Reserve, at Hill AFB, Utah. *(Ben Knowles via the Mick Roth Collection)*

Resplendent in its red, white, and blue paint scheme, this Thunderbird F-105B-10-RE (57-5782) was captured on its delivery flight on 25 January 1964. The pilot was Major David Pilton. Although its career as a demonstration aircraft lasted less than a year, there were at least three variations in the way the blue and white scallops were painted on the nose. *(Cradle of Aviation Museum via Ken Neubeck)*

Everybody celebrated the Bicentennial during 1976. The 141st TFS of the New Jersey Air National Guard painted up this F-105B-10-RE (57-5776) with special tail markings. *(Mick Roth Collection)*

The last F-105D-5-RE (58-1173) was used in various test programs. Here it is shown carrying a full load of 16 M117 750-pound bombs. Note the red markings on the air intakes, around the nose, and under the wings. *(San Diego Aerospace Museum Collection)*

Operation Look Alike painted all F-105s in an overall silver lacquer in order to seal the aircraft against moisture. Although it was not as attractive as the earlier natural metal finish, it afforded some protection for the electronics against the elements. This F-105D-20-RE (61-0143) is missing the TAC badge that was normally in the middle of the lightning bolt on the vertical stabilizer. *(Mick Roth Collection)*

The second unit outside the United States to convert to the F-105 was the 49th TFW at Spangdahlem AB, Germany, during October 1961. This F-105D-10-RE (60-0513) still retains its natural metal finish, which was replaced with the Look Alike silver in the summer of 1962. *(Roger Warren via the Larry Davis Collection)*

F-105D-5-RE (59-1737) taxies at Korat RTAFB with six M117 general-purpose bombs on the centerline and a LAU-3 rocket pod on each outer wing station. The aircraft was lost on 25 October 1967 when it collided with a C-123 while landing. The aircraft was assigned to the 469th TFS, and Major A. F. Britt was killed in the accident. *(Larsen/Remington via the Mick Roth Collection)*

"Blitzkreig" was an F-105D-25-RE (61-0208) that was lost on 19 November 1967. The aircraft was hit by a SAM over North Vietnam, killing Captain H. H. Klinck. *(Larsen/Remington via the Mick Roth Collection)*

This F-105D-10-RE (60-5375) was assigned to the 333rd TFS at Takhli but was flown by Colonel C. E. "Bud" Anderson, Commander of the 355th TFW. The aircraft carried the name "Old Crow II" on both intakes, and four-color striping on the nose, radar reflector, and pitot boom—one color for each of the squadrons assigned to the 355th. *(C.E. Anderson)*

"My Karma" was an F-105D-31-RE (62-4301) assigned to the 466th TFS, Air Force Reserve, at Hill AFB. The experimental wraparound "Europe I" paint scheme later evolved into the European One camouflage used extensively during the 1980s. Essentially this replaced the tan used in the Southeast Asian camouflage with a dark gray. *(Paul Minert via the Mick Roth Collection)*

Another experimental paint scheme was "Desert Fox," seen on F-105D-31-RE (62-0299) from the 466th TFS. This photo was taken at CFB Bagotville on 26 June 1983. *(Terry Panopalis)*

This Wild Weasel III F-105F-1-RE (63-8321) was assigned to the 357th TFS at Takhli in December 1968. A single AGM-45 Shrike is under the outer wing station, and a single M117 750-pound bomb is on the inboard station. *(Cradle of Aviation Museum via Ken Neubeck)*

A load of M-117 750-pound bombs awaits loading on an F-105F-1-RE (62-4351) from the 354th TFS at Takhli. The upper az-el homing antenna appears to have been replaced by a blank plate. *(Cradle of Aviation Museum via Ken Neubeck)*

A Wild Weasel III F-105F-1-RE (63-8312) assigned to Ryan's Raiders. Note the green and tan undersides. This aircraft was shot down by a SAM on 29 February 1968 over North Vietnam. The crew, Major C. J. Fitton, Jr. and Captain C. S. Harris, were both killed. *(Larsen/Remington via the Mick Roth Collection)*

An F-105G-1-RE (63-8306) from the 17th Wild Weasel Squadron. Note the tan paint covering the forward section of the ALQ-105 blister. Normally this was painted gray. *(John Huggins via the Mick Roth Collection)*

A Wild Weasel F-105G-1-RE (63-8300) from the 6010th WWS at Korat. The aircraft is carrying an AGM-45 Shrike on the outer pylon on the left wing, an AGM-78 Standard ARM on the right inboard pylon, and an ALQ-87 ECM pod on the right outboard pylon. The aircraft is missing the QRC-380/ALQ-105 blisters on the fuselage. *(Don Logan via the Mick Roth Collection)*

... [Air Force Headquarters] approved 13 tests to be conducted without delay. The AN/DPN-61 radar homing set developed specifically for the F-105D fighter aircraft by Bendix Corp is listed as one of the 13 items to be tested. F-105D tail number 138 arrived at Eglin 28 Oct with DPN-61 installed. Testing is to begin immediately under the Wild Weasel program. APGC [Air Proving Ground Center] and TAWC have agreed to add this aircraft and the Vector equipped F-105D number 291 to Wild Weasel I. Test order and project directive are being amended to provide evaluation of these two F-105D aircraft equipped with the two different [RHAW] devices. A separate Wild Weasel IA report will be made. Accordingly, a separate DPN-61 test and report as originally included in the 13 tests is not contemplated at this time. Request concurrence.[153]

The change was approved almost immediately, leaving only the F-105F under Wild Weasel II. However, by the time the aircraft had been modified by SMAMA and delivered to Eglin in early December, the APS-107 had been dropped from consideration as a Wild Weasel system for the F-105. The Wild Weasel II project quietly disappeared, but its component parts survived. Although the APS-107B (X-1) and ALQ-51 were installed on the same F-105F and tested concurrently, they were assigned separate APGC project numbers and reported on individually. The APS-107B (X-1) evaluation was designated APGC Project 0435Y, while compatibility testing of the ALQ-51 on the F-105F was added to Project 0420Y (APR-25, APR-26, and ALQ-51 testing on the RF-101C).[154]

The internal APS-107B (X-1) proved to be a considerable improvement over the pod-mounted APS-107 (XA-1), but it was not yet ready for operational use. APGC recommended that the APS-107 be considered for use in Southeast Asia after correction of two deficiencies. Although no APS-107 variants were ever used operationally in the F-105, the first production version, the APS-107A, was adopted as the standard RHAW system for the F-4D and also used in the Navy's A-6B Mod 0 Iron Hand aircraft. The ALQ-51 was rejected as a potential F-105 ECM system due primarily to lack of space in the airframe for its components, but tests in the RF-101C were successful. The Navy later provided enough ALQ-51s to equip PACAF's Voodoos.

Wild Weasel IA

The DPN-61 set used in 61-0138 was somewhat unusual in that it was based on an antiradiation missile guidance system originally developed by Bendix for the Crossbow missile. The Air Force had planned to buy the DPN-61 for the F-105D in 1960, and in June of that year issued a letter contract to Bendix for an initial buy of 85 sets. The following December, however, the F-105 project office reevaluated the requirement for a radar-busting capability in the F-105D and for a number of reasons—not the least of which was to lower the cost of the expensive Thunderchief—recommended that development of the DPN-61 be terminated. At the time, a radar-busting system fell into the nice-to-have-just-in-case-we-need-it category; in September 1965 it was a top-priority necessity. The Air Force approved the DPN-61 for testing as a stand-alone homing and potential acquisition system for the Shrike missile on the F-105D, and in mid-October it was ready for installation and testing. By then, however, the Air Force was less interested in an offensive system that could only detect radars in a narrow cone ahead of the aircraft than one that offered both homing and warning capabilities. At about the same time the Maxson fin cap warning receiver became available for testing and the Air Force decided to combine both receivers into a RHAW suite.[155]

The DPN-61 was a monopulse receiver that operated in the 2-11 GHz frequency range. Four triangular log periodic antennas, mounted around the nose just aft of the radome, covered a 60° conical sector centered on the aircraft's boresight. Homing displays were presented on a cathode ray tube and threat identification via a light panel, both of which were shared with the Maxson warning receiver. The Maxson receiver used one omni and four broad-band log periodic antennas mounted in the fin cap, along with their associated preamplifiers, to provide 360° azimuth and ±30° elevation coverage of threats in the 2–11-GHz frequency range. Both systems presented audio tones to the pilot.[156]

The APR-25 RHAW set installed in 62-4291 was identical to the Vector IV used in the F-100F Wild Weasel. The only differences in the installation involved mounting hardware and fairings. An antenna fairing was mounted under the nose to house the system's two 3-inch-diameter forward antennas. A simple crystal video receiver for each antenna was mounted near the fairing but within the fuselage. Two rear antennas, identical to those under the nose, were mounted on the trailing edge of the vertical stabilizer above the rudder, with their receivers in the fin cap. Cabling was installed to tie the nose and tail antenna installations to a signal analyzer located in the nose. The analyzer—the heart of the APR-25—was connected to a 4.5-by-5.75-inch combined azimuth indicator scope and threat display panel mounted atop the instrument panel shroud, to the right of the optical sight. The APR-25's frequency coverage was 2.4–3.6, 4.9–5.1, and 7–11 GHz. Received signals were analyzed by frequency, pulse repetition frequency (PRF), and scan rate with threat indications presented as coded strobes on the azimuth indicator, threat light illumination, and audio tones.[157]

The Wild Weasel IA test, which was completed in late December 1965, demonstrated that for warning purposes the APR-25 was superior to the DPN-61/Maxon system. The test also clearly demonstrated that RHAW systems were less effective in single-seat tactical aircraft than in two-seat aircraft, where a trained electronic warfare officer (EWO) could devote his full attention to the equipment. The DPN-61/Maxon system was by far the more difficult to use—the pilot had to deal with five separate control or indicator units located in the right console and to the left and right of the optical sight. Furthermore, the system could not operate in homing and warning modes simultaneously. This overloaded the pilots and reportedly led to one pilot becoming so task-saturated that he nearly flew into the Gulf of Mexico while setting up for a homing run on an Eglin threat simulator.[158]

TAWC and APGC concluded that the DPN-61/Maxon and APR-25 systems, when employed for homing on a target radar in dense signal environments, required more attention to signal identification and interpretation than could be given by an F-105D pilot. The project's final report recommended that "the F-105D aircraft, as equipped during this test, not be considered to have an effective radar homing capability and they are not recommended for operational employment as pathfinder or hunter-killer aircraft in a multiple signal environment."[159]

Wild Weasel III

In late November the F-100F Wild Weasels deployed to Thailand and by the end of the year had achieved their first SAM site kill, validating the electronic hunter-killer concept. Wild Weasel I experience to date also underscored the basic mismatch in performance between the F-100F hunter and the accompanying F-105D killers: the faster Thunderchiefs were forced to weave continuously behind the Wild Weasel in order to remain in position. It was clear that an F-105 Weasel was required.

Based on the success of Wild Weasel I, the results of the Wild Weasel IA evaluation, and the unsuitability of the APS-107 as a stand-alone Wild Weasel system, the Air Staff directed that a third F-105 Wild Weasel project be undertaken. Wild Weasel III would employ the proven Wild Weasel I avionics (APR-25, APR-26, and IR-133) in seven F-105Fs, five of which were to deploy to Southeast Asia in April 1966. The other two would remain at Eglin AFB for use in training additional Wild Weasel crews and testing modified or new equipment.

The APR-25 components used for Wild Weasel III differed from those in the Wild Weasel I F-100F only in the addition of a front cockpit threat display unit.

The IR-133 consisted of a receiver, three flush-mounted spiral antennas, and a cathode ray tube display and associated controls for the EWO. Two of the antennas, used for direc-

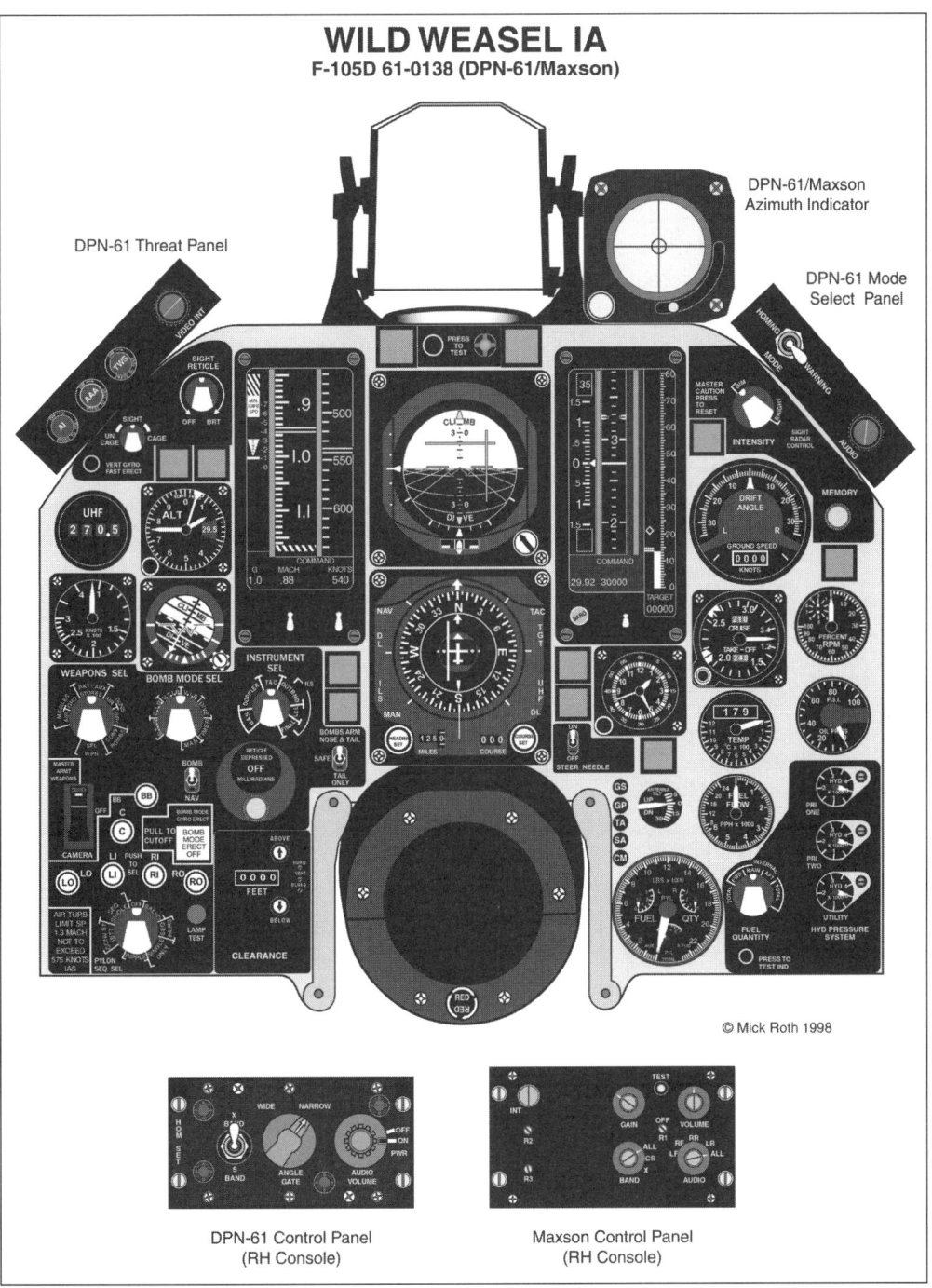

The DPN-61 homing and Maxson warning sets were installed in a single F-105D-20-RE (61-0138) for testing during Wild Weasel IA. Homing displays were presented on the CRT to the right of the optical sight, and threat identification via a light panel to the left of the sight, both of which were shared with the Maxson warning receiver. *(©1998 Mick Roth)*

tion finding, would be mounted on either side of the fuselage. The third antenna, used for omnidirectional search and signal analysis, would be mounted under the nose. The IR-133 was almost twice as sensitive as the Vector, allowing it to detect signals at greater range. Its scope display would allow the EWO to view all radar activity in the 2–4 GHz band, the region in which the Fan Song B and AAA radars such as the Fire Can operated, and simultaneously examine the pulse and scan characteristics of each signal as well as determine its frequency. The system also had a direction-finding (DF) mode in which a selected signal could be alternately sampled by the left and right antennas and displayed as a pair of vertical lines representing signal amplitude.[160]

The APR-26 launch warning set, which consisted of a blade antenna, a receiver, and cockpit controls and indicator lights, intercepted missile guidance signals transmitted by the SA-2's Fan Song radar. The system did not analyze the content of the guidance signal; it just looked for the presence of a 700–850 MHz signal within a specific PRF range and above a minimum signal strength threshold. Initial reception of such a signal was treated as a missile guidance activity condition. Detection of an abrupt power increase in the signal triggered a launch warning.[161]

The F-105Fs could carry a larger weapons load than the F-100Fs and would also be equipped to carry the standoff AGM-45 Shrike antiradiation missile. The Wild Weasel I aircraft originally had no Shrike capability and were not field-modified to carry the missile until April 1966.

On 5 January 1966, representatives from SMAMA, Republic, and ATI met at McClellan AFB to define the requirements for the Wild Weasel III F-105F and plan a quick-reaction modification effort that would deliver the seven aircraft in the shortest time possible. In order to expedite the modification process, the planners decided essentially to duplicate the proven F-100F installation.

For the next two weeks, ATI's engineers focused on adapting the Wild Weasel I equipment installation to the larger, more complex F-105F. In a 21 January meeting, ATI presented its technical data to Republic's Wild Weasel III project managers and emphasized the critical nature of certain design elements in the F-105 avionics installation. To prevent degradation of APR-25 and IR-133 sensitivity, low-capacitance coaxial cable was required for all video lines, with a minimum number of segments, and only low-loss straight connectors were to be used rather than right-angle connectors. Balanced cable runs were required to avoid DF errors.[162]

When SMAMA and Republic technicians began work on the prototype F-105F (62-4416), problems arose. Each aircraft required about 1,000 feet of coaxial video cable, and the type that ATI had specified proved to be in short supply. Republic was unable to obtain enough of it to complete even the prototype aircraft. Pressed by the tight schedule and unable to obtain coaxial cable at short notice, Republic proposed to SMAMA that shielded wire be substituted in the APR-25 installation. Before making a decision, SMAMA submitted Republic's proposal to ATI for analysis. Since the cable lengths and shielded wire specifications were as yet undetermined, ATI could only respond that while sensitivity would be degraded, they could not calculate the magnitude of the loss. Unable to quantify the effects of the wiring change and anxious to avoid delays, SMAMA approved the proposal.[163]

The prototype F-105F Weasel, which was designated the systems test aircraft, emerged from SMAMA's modification shops on 3 February. Because of the urgency of project, no time was available for ATI to perform a detailed system checkout and sensitivity test before the aircraft left McClellan for Eglin AFB 2 days later. By this time, work had begun on the next three aircraft (63-8298, 63-8302, and 63-8330). These were followed into SMAMA's shops on 15 February by the last three F-105Fs (63-8262, 63-8273, and 63-8286).[164]

Operational testing of the prototype, which began at Eglin on 9 February, quickly stumbled; the Weasel equipment, which had performed well in the F-100F, now exhibited severe sensitivity and accuracy anomalies. TAWC, which had been assigned responsibility for conducting the evaluation, called a halt after only a few sorties had been flown. When an inspection of the Weasel installation revealed a variety of engineering and quality defects, TAWC decided to defer testing until the second Weasel F-105F was delivered in late February.[165]

Back at McClellan, work on the next three Weasels was nearing completion. On 17 February, ATI was at last able to perform a quick sensitivity check on one of the aircraft. The results confirmed ATI's doubts about the APR-25 installation; the system's sensitivity appeared to be about 10 dB less than it should have been. On 21 February, ATI notified SMAMA and Republic of the degraded sensitivity and identified the cause as the use of shielded wire in lieu of low-capacitance video cable. However, because of the urgency of the Wild Weasel program, SMAMA was unwilling to delay delivery of the F-105s and took no action. ATI performed the first comprehensive sensitivity test of the Wild Weasel III installation on 24 February and measured a 15-dB signal loss, equating to an 82% reduction in reception range. Not only was the system's range reduced, but for signals originating near the aircraft's 3- and 9-o'clock positions, DF accuracy was degraded by approximately 18°. IR-133 homing information was also inaccurate, a consequence of the unequal number of right-angle connectors used in the right and left side cable installations.[166] The next day, ATI reported that the F-105F Wild Weasel equipment installation was unacceptable from the standpoint of sensitivity.[167]

While ATI was confirming the wiring deficiency in the Wild Weasel III installation, SMAMA and Republic were completing the F-105 modifications. On 23 February, the second Wild Weasel III F-105F (63-8302) arrived at Eglin. Before restarting the evaluation, TAWC inspected the aircraft and was disconcerted to find not only numerous deficiencies similar to those seen in the Wild Weasel III prototype but missing APR-25 and IR-133 components as well. After 63-8298 and 63-8330 arrived at Eglin two days later, TAWC's inspectors descended upon them and found similar problems. TAWC again postponed the evaluation and notified SMAMA, in a blunt 28 February message, that significant quality control and engineering deficiencies were present in every Wild Weasel III F-105F delivered to date.[168]

In response to TAWC's message, SMAMA recalled the first four aircraft and quickly assembled a Wild Weasel III Rework Team of SMAMA, TAWC, Systems Engineering Group, ATI, and Republic engineers to investigate and correct the reported defects. On 1 March, the rework team began inspecting and checking out the equipment installations of the three F-105s still at McClellan. When the Eglin aircraft arrived back at McClellan, the rework team not only verified TAWC's findings but uncovered other problems as well. Many were common to all seven aircraft, while others occurred in only two or three. Some, the result of equipment shortages, could not immediately be rectified; one aircraft had only the APR-26 installed and two others lacked IR-133 receivers and indicators. A mid-March rework team report listed 30 deficiencies in the Wild Weasel III aircraft, including short circuits, incorrect wiring, mismatched connectors, and electrical imbalances.[169]

By 3 March the dimensions of the problem were clear, and that afternoon Air Force, Republic, and ATI managers met to decide what corrective action was required and how best to proceed. It was apparent that the key issue was the APR-25 system's degraded sensitivity. To correct this deficiency the RHAW wiring installations would have to be restrung with video cable, a task that SMAMA engineers estimated would alone require some 300 man hours per aircraft to complete. There was, however, no realistic alternative; regardless of fixes to the other deficiencies, the APR-25 would never perform acceptably with the existing wiring. The group formulated a rework plan for the seven F-105Fs, and, since sufficient video cable was now available, work was in progress by that evening. Working around the clock, the team completed the first reworked Weasel (62-4416, which remained the dedicated test aircraft) on 9 March. The F-105F left for Eglin the next day, followed by the other six between 12 and 18 March.[170]

Testing was restarted on 14 March. Unlike the Eglin Wild Weasel I evaluation in which all aircraft and aircrews participated in simultaneous testing and training, in Wild Weasel III one aircraft flew most of the test sorties while the other six were used primarily for training. Two crews were heavily involved with testing, while eight additional crews concentrated on training and tactics development for the upcoming deployment to Korat. All aircrews were sent to NAA's Long Beach, California, Special Projects facility for ground training and avionics familiarization in the simulator originally built for the Wild Weasel I project.

At about the same time, with the air war in Southeast Asia heating up, Air Force headquarters directed SMAMA to modify six more F-105Fs to the Wild Weasel III configuration for a total of 13 aircraft. Modification of these additional F-105Fs was to begin the first week in May, with all six aircraft completed by the end of the month.[171]

In late February the Air Force decided to include an additional system in the Wild Weasel III equipment suite. Even with the help of radar homing equipment to guide them, Wild Weasel I crews were finding it extremely difficult to visually locate North Vietnamese SAM sites. The F-105F Weasels had the same electronic systems, and their crews would face similar difficulties. ATI and American Electronic Laboratories (AEL) had developed azimuth-elevation (az-el) terminal homing systems designed to display the location of a transmitting SAM radar in the pilot's optical sight. The az-el systems used log periodic antennas (four for ATI's system and five for AEL's) spaced around the nose to intercept signals from radars located within a narrow cone ahead of the aircraft. The antennas were sampled sequentially and the signal amplitudes compared to establish a line of sight to the target radar. A small green dot was projected onto the optical sight combining glass, superimposed over the target. ATI's AE-100 consisted of homing antennas and a signal comparator connected to the IR-133 E–F-band receiver, while AEL's Pointer III, with its own E–G-band receiver, was entirely separate from the IR-133. After comparing the competing systems, the Air Force selected the Pointer III. The planned 1 April deployment of five Weasels to Korat RTAFB was pushed back a month to allow time for installation, testing, and aircrew training.[172]

On 20 March F-105F 63-8273 was flown from Eglin to Republic's plant at Farmingdale for prototype installation and checkout of the system. Between 27 March and 15 April, Pointer IIIs were installed in five of the remaining six Weasels at Eglin, and flight testing of the system began during the first week of April aboard 62-4416. AEL's az-el system performed so poorly in flights against Eglin's simulated threat radars that the Air Force, in mid-April, decided to install and test ATI's az-el system in the seventh F-105F. The AE-100 did not perform flawlessly, but it was a usable system and a considerable improvement over the Pointer III. On 22 April the Air Force decided to replace the Pointer IIIs in the first six Weasels with AE-100s. This again delayed the Korat deployment, which was rescheduled for 22 May.[173]

Another potential Weasel system became available in late April and was given a quick-look test in 62-4416 before the end of the month. SEE-SAMS (sense, exploit, and evade surface-to-air missile system) was designed to provide more detailed information on the degree of threat posed by an SA-2 site than was available from the APR-25, APR-26, and IR-133. Like the APR-25/26, SEE-SAMS provided warnings when the aircraft was illuminated by a Fan Song operating in high PRF mode and when missile guidance signals were detected. While these warnings were indicative of an SA-2 launch, other aircraft near the SAM site's target also received the same signals. SEE-SAMS, by analyzing intercepted signals to determine the aircraft's position relative to the center of the Fan Song's horizontal scan sector, was designed to alert the crew of the targeted aircraft. An aircraft had to be within the 10° by 10° common sector scanned by the Fan Song's azimuth and elevation beams before it could be engaged, and according to intelligence information available at the time, Soviet doctrine required that the target aircraft actually be centered in the sector before missiles were launched. Only when SEE-SAMS determined that the aircraft was centered in the Fan Song's crosshairs did a warning light alert the crew that they were likely to be the recipients of a Guideline.[174]

Development of SEE-SAMS had begun at APGC the previous fall, and by mid-November a breadboard model of the system had been tested successfully aboard C-131 and T-39 aircraft against Eglin's SADS-I Fan Song simulator. Based on the success of the Eglin tests, Systems Command requested $150,000 from Air Force Headquarters to procure approximately 20 SEE-SAMS sets for evaluation at Eglin and in Vietnam. NAA was awarded the contract to build the sets under QRC-317.[175]

The first set was delivered to Eglin on 18 April and installed in the systems' test aircraft. A shakedown test sortie, flown against APGC's QRC-207 Fan Song simulator on 23 April, yielded disappointing results: the set's false indications far outnumbered the correct ones. After additional testing against the Flintstone/Big Pebble Fan Song simulator operated by Sanders Associates near Nashua, New Hampshire, revealed additional deficiencies the system was removed from the aircraft and returned to NAA for redesign.[176]

The Air Force also conducted a quick-look evaluation of a potential APR-26 replacement in April. An HRB-Singer 934-1B missile warning receiver was installed in 62-4416 and test flown at the Sanders facility, which had a Fan Song missile guidance simulator not available at Eglin. The 934-1B differed from the APR-26 in that it analyzed the modulation characteristics of the C-band guidance signal to differentiate between SA-2 missile activity and missile launch modes, while the APR-26 simply looked for an abrupt amplitude increase. The HRB-Singer set performed well, but the Air Force was already committed to a large APR-26 procurement and saw no compelling reason to buy another system to perform the same function. Only after the Wild Weasel III F-105s were in combat was it learned that the APR-26's design was based on possibly faulty intelligence regarding the amplitude increase. This led to numerous incidents of false lower threat-level "activity" indications when "missile launch" should have been displayed. The APR-26 was later modified to analyze the guidance signal and the improved sets redesignated APR-37.[177]

On 22 May 1966, five Wild Weasel F-105s and eight aircrews (Wild Weasel III-1) deployed to Korat, where they were assigned to the 13th TFS, 388th TFW. On 3 June the first theater orientation mission was flown in Route Package 1, led by a Wild Weasel I aircrew in an F-100F, and by 9 June Wild Weasel F-105Fs were flying into Route Package 6.

By late June NAA had produced seven redesigned SEE-SAMS sets, called SEE-SAMS B, and four were sent to Korat for combat testing. This was the first of numerous equipment modifications applied to Wild Weasel F-105s. Some system improvements were introduced incrementally as new batches of Wild Weasels were modified at SMAMA, while others were applied to only a few selected aircraft. By the time the F-105 Weasel equipment was standardized in the F-105G, more than 15 F-105F configurations had seen combat. Some of the differences were relatively minor, but collectively they represented a source of confusion to aircrew who had to deal with varying system configurations and capabilities. The many configurations also represented a maintenance and logistics nightmare.[178]

When the Wild Weasel III-1 F-105Fs deployed to Korat in late May, the last of eight Wild Weasel III-2 aircrews for the six follow-on aircraft was completing ground training at NAA's Long Beach facility. Flight training followed at Nellis AFB, in the just-established Wild Weasel training program. Because Wild Weasel operating concepts and tactics were still very much in the evolutionary stage and intelligence on the threat was thin, the training provided to the aircrews in the first few classes was at best rudimentary. No threat radar simulators were then available at Nellis, so flight training consisted mainly of sorties designed to familiarize the aircrews with the Weasel F-105F. The training was geared more to the EWOs—most of whom were used to flying in B-52s, not fighters—than the high-time F-105 pilots.[179]

The six F-105Fs and eight Wild Weasel III-2 aircrews were to deploy to Korat at the end of June, at which time the Wild Weasel III-1 detachment would transfer to Takhli. The Wild

Weasel I F-100F crews, who were winding down operations at Korat, would provide initial theater orientation and training for Wild Weasel III-2 before returning to the United States in late July. At virtually the last minute, however, the Air Force decided that moving the 388th TFW Weasels to a different base would be too disruptive of Wild Weasel operations and canceled the transfer. By then an advance team of two Wild Weasel III-2 aircrews had already arrived at Korat. Two days later, on 3 July, special orders arrived transferring them to the 355th TFW. The six F-105Fs and the remaining aircrews arrived at Takhli on 4 July. Unlike the 388th TFW, which assigned all its Weasels to the 13th TFS, the 355th TFW distributed two F-105Fs to each of its three squadrons. Assisted by a few 388th Wild Weasel III-1 aircrews temporarily detailed to Takhli, Wild Weasel III-2 began flying combat missions within a week of their arrival.[180]

The Weasels of both wings flew two basic types of missions—strike escort/Iron Hand, and hunter-killer. In the former, the goal was to attack and destroy North Vietnamese SAMs and radar-directed AAA that directly threatened the strike force. An Iron Hand flight of one Weasel F-105F and three F-105Ds typically preceded the strike force into the target area and remained there until after the strike aircraft withdrew. Once the strike aircraft were out of harm's way, the Weasels would switch to the hunter-killer role, fuel and unexpended weapons permitting. In the hunter-killer role the Weasels were not tied to a strike force and could search for and attack any sites located within approved armed reconnaissance areas.[181]

The weapons and tactics used were initially similar to those of the F-100F Weasels. The Shrike missile provided a standoff capability, but for a number of reasons it was not very effective. It was relatively difficult to employ due to its limited range and restrictive launch window, and this was exacerbated by the inability of the Weasel equipment to determine the range to a target. The Shrike also had a small warhead with limited destructive capability. Furthermore, once the radar operators became aware that an antiradiation weapon was being used against them, they began to limit their transmissions to frustrate attacks.

F-105Ds were often used as the "killer" part of hunter-killer teams. Here an F-105D-5-RE (58-1161) from the 469th TFS carries an AGM-45 Shrike missile in May 1966. *(U.S. Air Force/DVIC)*

Even after a successful Shrike attack the Weasels and accompanying F-105Ds often followed up with a low-altitude attack employing the same close-in weapons used by their F-100 predecessors: 2.75-inch rockets and 20-mm cannon. These attacks, combined with their increased exposure time in high-threat areas, took a heavy toll on the Weasels. By mid-August only 5 of the 11 Weasels deployed to Southeast Asia were still operational: one 388th and four 355th F-105Fs had been lost in combat and another was out of service due to battle damage.[182]

In July USAF headquarters had directed SMAMA to modify 18 additional F-105Fs as Wild Weasels, with modifications to begin on 1 September. The new Weasels would have the E–G-band ER-142 panoramic receiver in place of the IR-133 and qualified versions of the APR-25/26 to replace the prequalified sets used on the first 13 aircraft. The ER-142 was functionally similar to the IR-133, but where the IR-133 had a single 2–4 GHz scope the ER-142 had two panoramic scopes, one covering 2–4 GHz and the other 4–6 GHz.[183]

On 11 August 1966, in response to the heavy Weasel losses over North Vietnam, USAF headquarters asked SMAMA to accelerate the modification and delivery of the 18 new Wild Weasel F-105Fs, and in November added 36 more. The number was increased again in June 1967 when USAF headquarters directed that a further 19 F-105Fs be modified as Weasels, bringing the total to 86. The last of these were completed in early 1969.[184]

The heavy losses and increasingly capable North Vietnamese air defense system led to a reassessment of Wild Weasel employment and tactics. By mid-1967 the destruction of SAM sites by strike-escort Iron Hand aircraft was no longer considered the primary goal—it was now threat radar suppression. While destruction was the ultimate form of suppression, if Fan Song or Fire Can operators could be intimidated into shutting down their radars or perhaps even not coming up at all, that was fine too. SA-2 site destruction was by no means ruled out, but it was subordinated to protection of the strike force.[185]

This 355th TFW F-105F-1-RE (63-8277) carries an AGM-45 Shrike on the outer pylon and a CBU-24 cluster bomb on the inboard pylon. Cluster bombs replaced the LAU-3 rocket pods used by the F-100F and early F-105F Wild Weasels. *(Paul Chesley via Larry Davis via the Mick Roth Collection)*

In 1966 the Air Force began investigating methods of equipping the Wild Weasels with self-protection ECM equipment that did not occupy external stores stations as did ECM pods. A blister-mounted system was developed under QRC-321 and later updated to the QRC-380 configuration. F-105F-1-RE (63-8334) was the prototype aircraft for the QRC-380 internal ECM package modification. QRC-380 was later redesignated ALQ-105 and was eventually installed on all F-105Gs. This aircraft also has a small, unidentified fairing under the fuselage just forward of the nose landing gear. *(Terry Panopalis Collection)*

With more Wild Weasel F-105Fs available, the composition of Iron Hand flights changed from one F-105F and three F-105Ds to two elements consisting of an F-105F and an F-105D or two F-105Fs. One flight typically preceded the strike force into the target area by about five to seven minutes (this was reduced to around one minute in early 1968), while another flew toward the rear of the strike formation. Weasel aircrews varied their target area tactics from mission to mission. As one EWO put it: "You can do anything once. You might be able to do it twice. But you certainly can never do it three times, because the third time they'd have it figured out. . . . Our forte was doing the unexpected . . . because the minute you did the expected you got killed." The SAM operators varied their tactics as well, resulting in a continual series of moves and countermoves. The trend on the North Vietnamese side was toward minimal transmission times, an increasingly dense network of SAM sites positioned to maximize mutual support and frustrate Shrike attacks, and greater air defense system integration.[186]

The armament carried by the Weasels also changed; 2.75-inch rockets and 20-mm cannon fire were replaced by CBU-24s and 500- or 750-pound bombs as the weapons of choice to follow up a Shrike attack. These weapons were not only more destructive, but they could be delivered from higher altitudes and didn't require the aircraft to overfly the site at low altitude. The 1967 edition of the 388th TFW F-105 Combat Tactics manual offered the following guidance to Wild Weasel crews:

> **. . . Unless the flight leader knows exactly where the SAM/Fire Can is located and can plan on a surprise attack, it is generally good practice to initiate an attack with a Shrike launch for two reasons:**

F-105Fs assigned to the 66th Fighter Weapons Squadron of the 57th FWW at Nellis AFB, Nevada, were used to train Wild Weasel aircrews as well as test new Weasel equipment. Note the weapons bay fuel tank in the lowered position under this F-105G-1-RE (62-4442), photographed in December 1971. The 57th traditionally used a yellow and black checkerboard band around the vertical stabilizer. *(Mick Roth)*

> a. It places a tracking radar site on the defensive and buys time for the flight to penetrate the critical zone between the enemy's maximum and minimum effective range.
>
> b. The Shrike impact furnishes an easily observed reference point for the flight leader to use in directing the attack.
>
> ... The limited destructive power of the Shrike missile must be augmented by high-explosive weapons to attain a reasonable assurance of SAM/Fire Can site destruction. CBU-24 munitions give the necessary area coverage and possess sufficient destructive power to neutralize radar and fire control systems within the site. However, high-explosive bombs are required for missile/gun destruction. Current practice is to have the accompanying F-105D aircraft carry 500 lb bombs.[187]

In the spring of 1967 Air Force Logistics Command requested that Wild Weasel F-105Fs, with their special avionics and unique mission, be redesignated EF-105Fs. Air Force headquarters responded in early June, disapproving the redesignation.[188]

New electronic equipment and modifications to existing systems were continually under development for the Weasels. Loral developed a SEE-SAMS-type of receiver, which was tested in late 1966. Although the unit never went into production, Loral was awarded a contract to upgrade the NAA-produced SEE-SAMS under QRC-317A. Seven sets were installed in Weasel F-105Fs during the summer of 1967. The QRC-317A was later upgraded and produced in quantity as the ALR-31, which was retrofitted into a number of ER-142 F-105Fs by TCTO 1F-105F-540 beginning in July 1968 and was an integral part of the F-105G configuration.[189]

Also in 1966 the Air Force began investigating methods of equipping the Wild Weasels with self-protection ECM equipment that did not steal one or more external stores stations, as did the QRC-160 (ALQ-71 and -87) and QRC-335 (ALQ-101) jamming pods. The first of these projects was the Sanders QRC-301, in which a version of the Navy's ALQ-51 deception jammer was scabbed onto the outboard pylons of an F-105F. The project was followed by QRC-321—

The two antiradiation missiles used in Southeast Asia were the AGM-45 Shrike (on the outer pylon) and the AGM-78 Standard ARM (on the inboard pylon). Note the size difference between the missiles. The dark, rectangular panels behind the radome of each missile are fuze antennas. *(via Larry Davis via the Mick Roth Collection)*

a fuselage blister-mounted version of the Westinghouse QRC-288 deception jamming pod—and QRC-380 with repackaged QRC-335 jammer components mounted in the QRC-321 blister. QRC-380, later redesignated ALQ-105, was eventually produced for all F-105Gs.[190]

In 1967 improvements to the APR-25 and APR-26 were developed and entered testing at Nellis and Eglin. The APR-25 improvement, known as Spot SAM, incorporated the SEE-SAMS technique of relating the degree of threat to the aircraft's position in the Fan Song scan sector. A separate modification provided the capability to correlate a C-band missile guidance signal received by the APR-26 to a specific E–F-band signal displayed on the APR-25 azimuth indicator. The APR-26 was upgraded to analyze the content of Fan Song missile guidance signals rather than just looking at power levels. The upgraded systems were later produced as the APR-36 and APR-37, respectively.[191]

Other systems developed and tested in 1967–68 that were later incorporated into the F-105F/G were the Borders QRC-373/ALT-34 acquisition radar jamming system (four separate transmitters operating in the 104–900 MHz frequency range) and the ER-168/APR-35 panoramic and analysis receiving system, which replaced the ER-142.[192]

Not all systems tested during this period were as successful . . .

The success of the Wild Weasels in Southeast Asia had forced the North Vietnamese to change their SAM and AAA radar tactics. Threat radars were radiating less and responding to Weasel countermoves by shutting down. The increasing sophistication and integration of the air defense system made more use of VHF and UHF (70–900 MHz) acquisition radars such as the Spoon Rest to pass information to weapon control radars, allowing them to radiate for a minimal amount of time. Because acquisition radars were normally located near—if not adjacent to—SAM sites, the Air Force began looking for a system to home on them. QRC-333 was the result. Two competitive receivers were tested on a few F-105Fs (possibly in Southeast Asia), but it never became a standard Weasel system.[193]

Another system that surfaced in 1967 and quickly disappeared was the QRC-339/ER-151 automatic RHAW system. Anecdotal evidence suggests that this was another attempt to develop a Weasel system usable in a single-seat fighter. Whatever the intent, the ER-151 never worked properly and its development was abandoned.[194]

Radar receiving systems of the period had two major failings: they were unable to determine range to an emitter or precisely locate the emitter without a close approach or overflight. The Air Force conducted a series of studies in an attempt to develop a ranging system, but none bore fruit until the APR-38 was developed for the F-4 Wild Weasel in the mid-1970s. In the fall of 1968 the Air Force tried to solve the target location problem with the LTV bistatic aided strike system (BASS). The system consisted of an EC-121 Rivet Top aircraft, with its suite of ELINT and DF receivers, and a transponder in a Wild Weasel F-105F. When a threat radar was detected by the EC-121, the onboard controller would plot its position. If the radar also illuminated the Weasel, it would trigger the transponder, theoretically displaying to the controller the F-105's position relative to the radar. The controller could then vector the Wild

Another view of an AGM-45 and AGM-78 under the wing of an F-105G. Development of the Standard ARM began in early 1966, and the AGM-78A (Mod 0), and a modified narrow-band Shrike seeker targeted against the E-F-band Fan Song was fielded in early 1968. The AGM-78B (Mod 1) missile, shown here, used a more sophisticated Maxson seeker with expanded frequency coverage and greater sensitivity. *(John Huggins via the Mick Roth Collection)*

The Weasels were still capable of carrying a normal bomb load. Here an F-105F-1-RE (63-8302) named "Half A Yard" from the 44th TFS carries six Mk 82 500-pound bombs on the centerline and one more under each outer wing station. The aircraft is on a mission into Laos during 1970. *(U.S. Air Force/DVIC)*

Weasel flight to the target radar. Unfortunately, BASS didn't work. The 44th TFS flew 10 test missions in Route Package 1 without success. According to the 44th TFS history for October 1968, "the first few [mission failures] involved communications problems and equipment checkout. The final missions pointed to a group of deficiencies in the system itself, any of which singularly could have caused failure. . . . On none of the BASS missions did the F-105 transponder signal display itself on the C-121 scope. The reasons for this are being investigated by the LTV people. It may be an engineering problem or it may be that BASS is a dream." Despite a modification of the system to a BASS II configuration, the problems were never resolved, and in late November BASS was terminated.[195]

Developmental work was not limited to Weasel avionics; the weapons used by the Weasels were also continually being improved.

The AGM-45 Shrike missile was originally developed by the Navy in the late 1950s to counter an E–F-band threat. After a U-2 was downed during the Cuban Missile Crisis by an improved version of the SA-2 that employed a G-band Fan Song, the Navy initiated the Emergency Shrike Effort to develop an alternate G-band guidance section for Shrike. During the Vietnam War, new guidance sections were developed with angle gating capability and a greater variety of frequency coverage options. Because the flash of Shrike's relatively small warhead could be difficult to spot under the less-than-optimal visual conditions that were the norm over North Vietnam, the Navy developed a phosphorous warhead which generated a white smoke cloud to mark the detonation point for a follow-up attack.

Despite the many improvements made to Shrike, it still suffered from a number of operational deficiencies, the more significant of which were its small warhead, vulnerability to

The nose area of an F-105G. The 20-mm cannon muzzle is in the center of the shark mouth. Note that the M61 fires directly across the 3-inch diameter APR-35 left DF antenna. Above and below the DF antenna are two of the four triangular az-el/homing antennas. Under the fuselage, just behind the radome, is the fairing that contained the two forward RHAW antennas (APR-25, APR-36, or ALR-46, depending on the time period) and the strike camera. The three APR-35 analysis receiver omnidirectional stub antennas are visible on the panel just ahead of the nose gear well. *(San Diego Aerospace Museum Collection)*

enemy emission control countermeasures, and limited range. (It is difficult to generalize about the Shrike's range since it was highly dependent upon the launch aircraft's speed and altitude, as well as the delivery mode. According to the 388th TFW's F-105 Combat Tactics manual, for launch aircraft altitudes of 6,000–9,000 feet and an airspeed of approximately 500 knots, the Shrike's range envelope was considered to be 3–13 miles for planning purposes. The Shrike's maximum range could significantly exceed 20 miles, especially when the toss delivery mode was used at high altitudes, but only at the expense of a much-decreased probability of kill.)[196]

In early 1966 General Dynamics began work on a missile designed to exceed the capabilities of Shrike in all areas. The AGM-78A (Mod 0) Standard ARM was an air-launched version of the ship-launched Standard Type 1A SAM with a new warhead and modified narrow-band Shrike seeker targeted against the E–F-band Fan Song. The Shrike seeker electronics were used in order to field the missile as quickly as possible; work began shortly thereafter on the AGM-78B (Mod 1) missile, which would use a more sophisticated Maxson seeker with expanded frequency coverage and greater sensitivity.

The Standard ARM offered many advantages over the Shrike: the explosive weight of its warhead was three times greater, it was about 0.5-Mach faster, its effective range was more than three times longer, and its guidance system was far more sophisticated.

While Shrike had to be launched into a narrow "basket" above the target with little tolerance for azimuth error, Standard ARM could be programmed to turn as much as 180° after launch before looking for the target radar. While shooting at a radar behind the aircraft was possible, the capability was not very useful from an operational standpoint since the turn

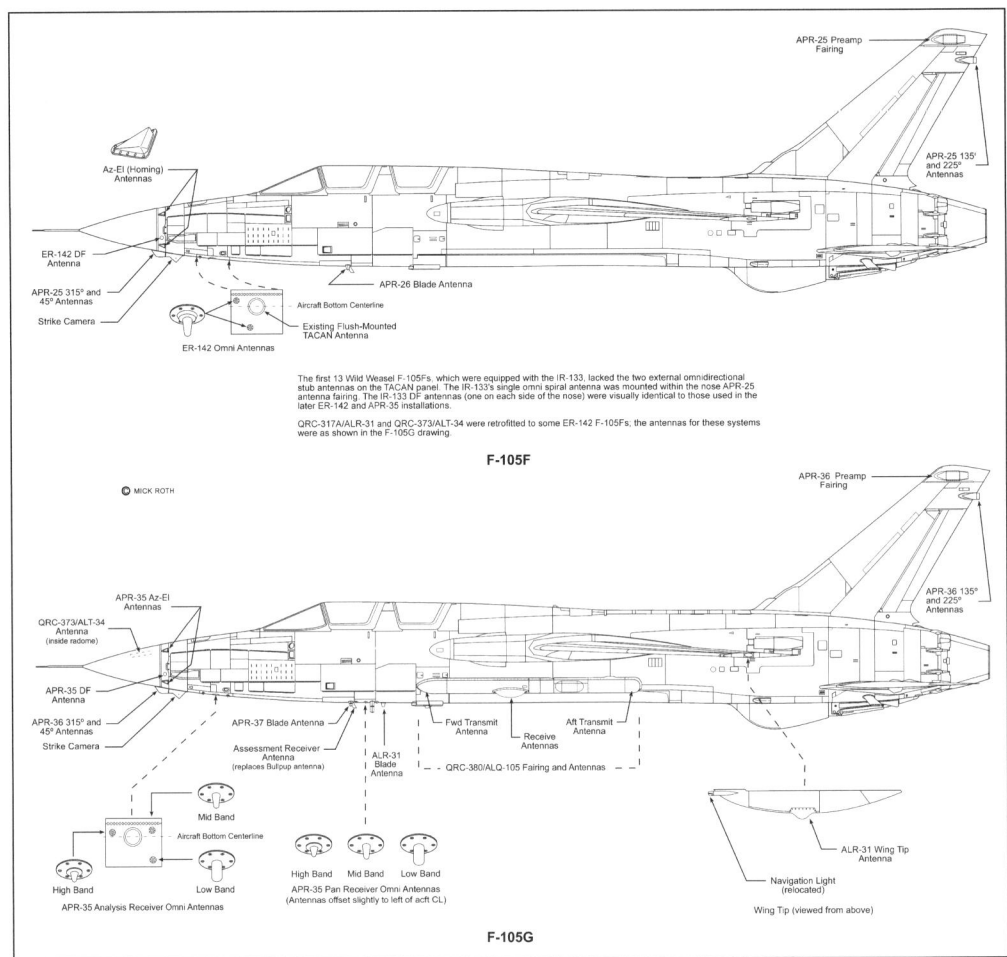

The differences in the antenna arrangement between the F-105F and F-105G are illustrated in this set of drawings. *(©1998 Mick Roth)*

used up most of the missile's energy, reducing its range drastically. Launches at targets off boresight but within the forward hemisphere did not incur nearly as large a range penalty and offered the tactical advantage of allowing the Weasel to shoot at a radar without distinctive maneuvers that advertised its intent to do so.

If the operators of a target radar shut it down while a Shrike was in flight, the missile "went stupid." Even if the radar came back up after a few seconds, there was a high probability that it would no longer be within the look angle of the seeker's fixed antenna. Standard ARM had a movable antenna and a memory circuit; if the target radar shut down, the seeker switched to rate-hold mode, in which the antenna (and the missile) continued to turn at a rate proportional to that existing when the signal was lost. This significantly improved the likelihood of reacquiring the target if it came up again, as well as lessening the miss distance if the radar switched off late in the missile's flight.

Although the AGM-78 represented a significant improvement over Shrike, it was by no means a flawless weapon. During testing in the fall of 1967, a number of missiles became unstable or failed to guide to their targets, and although several fixes were incorporated, reliability continued to be a problem after the AGM-78 entered operational service.

In September 1967, SMAMA began modifying the first of 14 Wild Weasel F-105Fs to be equipped with the Standard ARM Mod 0 weapon system. By early February 1968 eight of the Mod 0 F-105Fs were operational at Takhli, and on 9 March the first Air Force Standard ARM combat mission was flown (the first combat launch of a Standard ARM was from a Navy

The front cockpit of the F-105G was not terribly different from that of the standard F-105F. The ALR-46 scope is mounted on top of the instrument panel. The audio management panel, to the left of the scope, controlled the volume of the various warning tones generated by the ALR-46. Immediately above the ALR-46 scope is the pilot's ALR-31 threat light panel. After incorporation of the ALR-46, the ALR-31 system became redundant and it was eventually removed. *(Mick Roth)*

The aft cockpit of the F-105G bore little resemblance to that of the F-105F trainer. Note the APR-35 display panel at the lower center of the photo, immediately above the radar scope. The square, hooded scope at the left presented a panoramic view of signal activity in the three bands covered by the APR-35, while the smaller rectangular scope at the upper right of the panel presented a homing or signal analysis display, as selected by the EWO. *(Ben Knowles via the Mick Roth Collection)*

A-6B, on 6 March). Over the next few weeks, F-105Fs launched eight AGM-78As at SA-2 sites in the Hanoi area, but when President Lyndon Johnson halted the air war north of the 19th parallel on 1 April, Air Force Standard ARM missions ceased. The Navy continued to employ AGM-78As until Rolling Thunder ended on 1 November 1968, but no further Air Force combat launches occurred until early 1971.[197]

After 1 April the pattern of the air war over North Vietnam shifted from large raids in the heavily defended Red River delta to armed reconnaissance and interdiction of supply lines in the southern route packages, for which the Wild Weasels provided protection. There were few SAMs in this area in April, but during the next few months the North Vietnamese took advantage of the bombing halt further north and made a concerted effort to expand their SAM defenses into the area near the DMZ. The Weasels reverted from the suppression role to hunter-killers, attacking any identified SA-2 site—occupied or not—with general-purpose bombs. By September the North Vietnamese had abandoned their SAM expansion . . . for the time being.[198]

On 1 November 1968 President Johnson called a halt to Rolling Thunder, ending the bombing of North Vietnam. This did not put the Wild Weasels out of business, however, because interdiction of the so-called Ho Chi Minh Trail through Laos did not end—in fact it expanded. B-52s had been employed against supply lines in South Vietnam and Laos since 1965, but they were restricted from operating in high-threat areas. A buildup of air defenses on the southwestern border of North Vietnam that began in 1967 put them within range of SA-2s, requiring ECM support from EB-66s as well as Wild Weasel support. After 1 November the Weasels continued to fly in the suppression role in support of B-52 strikes, making shallow penetrations into North Vietnam when necessary to intimidate Fan Songs into shutting down.

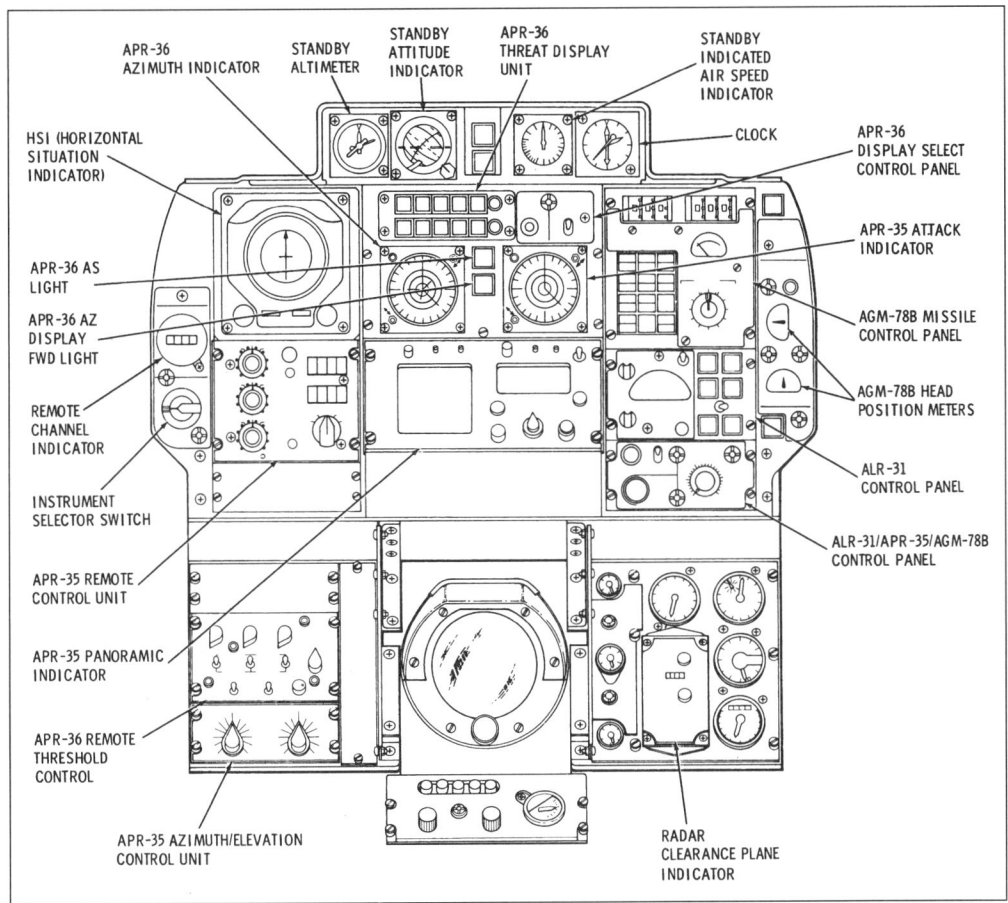

This is the original F-105G aft cockpit instrument panel, prior to the installation of the QRC-380/ALQ-105. The majority of panel space is dedicated to Wild Weasel displays and controls, with only minimal flight instrumentation remaining. *(U.S. Air Force)*

The Wild Weasels also supported manned reconnaissance flights, which continued after 1 November. After an RF-4C was downed over North Vietnam a few weeks after the bombing halt was imposed, the Weasels were authorized to suppress North Vietnamese AAA and SAMs threatening reconnaissance aircraft. As the North Vietnamese continued to expand their air defenses near the major supply routes into Laos, U.S. aircraft began flying protective reaction air strikes, which gradually increased in tempo.

In the fall of 1968, SMAMA began work on the first Standard ARM Mod 1 F-105Fs. Originally, only 16 aircraft were to receive this modification, which would allow them to employ the AGM-78B (and later the C) Mod 1 missiles. The upgrade replaced the Wild Weasel avionics, with the exception of the ALR-31 and QRC-373/ALT-34, and completely rearranged the EWO's instrument panel. The ER-142 panoramic receiver was replaced by the more sophisticated APR-35, which provided improved displays and frequency coverage from 2 to 10 GHz. The APR-25 and -26 were replaced by improved versions, redesignated APR-36 and -37, respectively. A second APR-36 azimuth indicator, referred to as an attack indicator, was placed on the EWO's instrument panel to be used for isolating and designating a specific signal for handoff to the Standard ARM. Although the attack indicator was used to display correlated information from multiple receiving systems, including the APR-36/37, it was a component of the APR-35 system. Finally, the Standard ARM Mod 1 missile control module was added, along with its associated controls and indicators, to interface the AGM-78 with the aircraft avionics. An associated modification carried out on the Mod 1 aircraft added a 14-channel tape recorder and BDA system. The BDA system consisted of a receiver in the

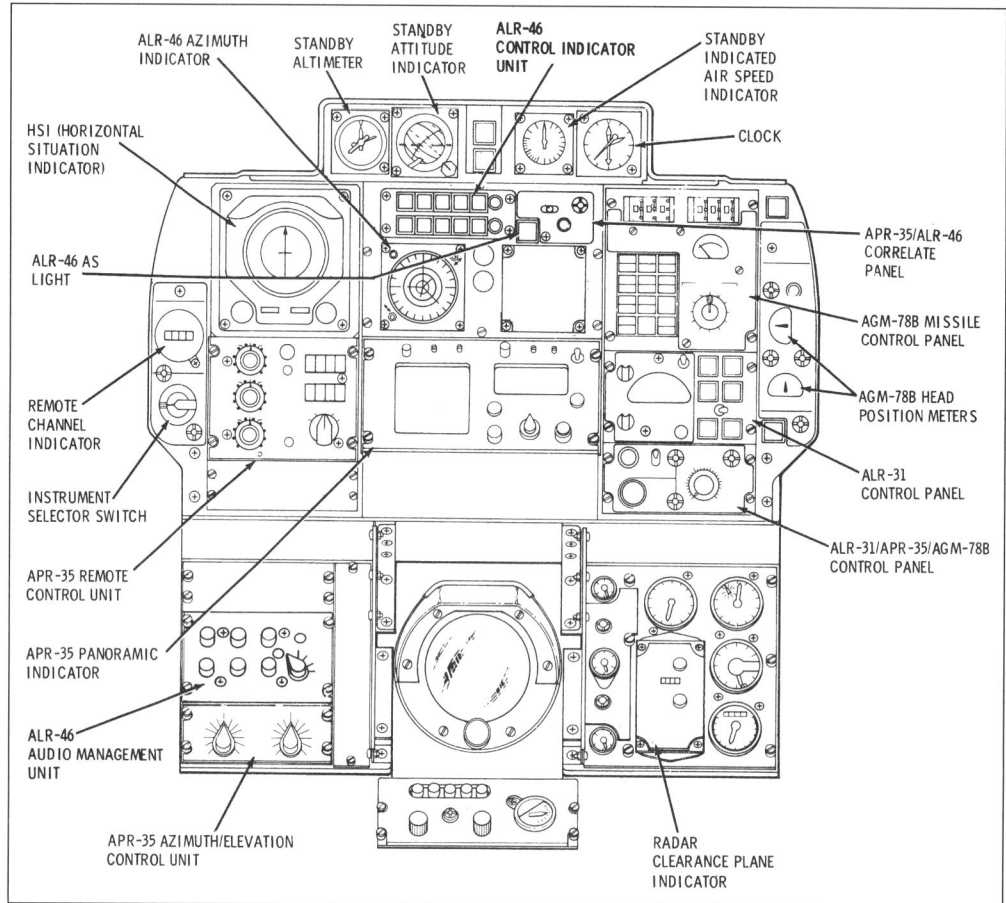

By 1974 the APR-36 and APR-37 had been replaced by the ALR-46. Note that the APR-35 attack scope has been removed from the aft cockpit. The ALR-31 control panel, immediately to the right of the APR-35 panoramic indicator in the aft cockpit, was later replaced by a blank panel. This drawing erroneously shows the radar clearance plane indicator on the aft cockpit lower right subpanel. In reality this location was used by the QRC-380/ALQ-105 control panel. *(U.S. Air Force)*

aircraft and a transmitter in the missile that repeated signals received by the seeker. The transmissions indicated whether the missile was homing or not, and postflight analysis of the recorded shutoff times of the missile transmitter and the target radar were used to assess the probability of a Fan Song kill.[199]

The first two Mod 1 aircraft rolled out of SMAMA's modification bays in January 1969, and work began on the third aircraft. Later that year the Air Force decided to upgrade its varied collection of F-105F Wild Weasel configurations to the Mod 1 baseline, and in October the Air Force Chief of Staff approved a redesignation of the modified aircraft as F-105Gs "to ensure that WILD WEASEL assets can be identified in appropriate programming documents and resources allocated accordingly." In 1971 12 non-Weasel F-105Fs were added to the modification program, bringing the total number of F-105Gs to 61.[200]

At least 14 F-105Gs had been completed when the aircraft's configuration was changed. The Air Force had been working on internal ECM for Wild Weasel F-105s since 1966, but it was not until mid-1970 that a suitable system became available. The QRC-380, later redesignated ALQ-105, used modified QRC-335/ALQ-101 components to provide both noise and deception jamming in the E–G-band frequency range. The installation (Class V modification 2079) resembled an ALQ-101 pod that had been split lengthwise and grafted to the F-105's fuselage below each wing. A number of the non-QRC-380-equipped F-105Gs saw combat in Southeast Asia and at least one was lost as a consequence of the lack of ECM

The first attempt at mounting a jamming system on the fuselage was the QRC-321—a fuselage blister-mounted version of the Westinghouse QRC-288 deception jamming pod. This was followed by QRC-380 with repackaged QRC-335 jammer components mounted in the QRC-321 blister. This system proved effective, and QRC-380 was later redesignated ALQ-105 and installed on all F-105Gs. Here the ALQ-105 is shown with its covers opened. *(Cradle of Aviation Museum via Ken Neubeck)*

protection. Since the Wild Weasels had been flying in relatively low-threat areas, they had become a bit complacent and no longer carried ECM pods on the pre-Mod 2079 aircraft. After one of the these aircraft was downed by an SA-2 in November 1971, the F-105Gs again began carrying pods, and an emergency program was initiated to certify the latest ECM pods for carriage on the F-105 Weasels until all aircraft were equipped with the QRC-380. Another modification, introduced in early 1972 and eventually incorporated in all F-105Gs, was the capability to carry two Shrikes on a single pylon using the ADU-315 (right pylon) and -316 (left pylon) dual-Shrike adapters developed by the Navy at China Lake, California. The adapters, each of which supported two LAU-34/A Mod 1 launchers, were cast magnesium units designed for suspension on the F-105's outboard pylons.[201]

Late 1970 saw the withdrawal of all F-105s from combat in Southeast Asia but for a small detachment Wild Weasels (12th TFS Det 1). The detachment was only in existence for a few months; the 6010th WWS replaced it in November 1970 and was itself replaced by the 17th WWS in December 1971.

During 1971 protective reaction strikes increased as the North Vietnamese continued to reinforce their air defenses near Ban Karai and Mu Gia passes, two of the major supply routes into Laos. In late December, U.S. aircraft flew more than 1,000 sorties during Proud Deep Alpha, a 5-day series of strikes against airfields and supply concentrations across

An F-105F-1-RE (62-4414) was used to test the dual-Shrike launcher at Eglin AFB prior to its being used in Southeast Asia. The missiles in this shot are inert rounds. Also shown in this photo is the lead-edge flap deflected slightly downward, and the original wingtip with the navigation light on the outer edge. *(Cradle of Aviation Museum via Ken Neubeck)*

southern North Vietnam. Navy and Air Force Iron Hand aircraft fired 51 Shrikes and 10 Standard ARMs at AAA, SAM, and GCI radars during the strikes, but the North Vietnamese managed to launch at least 45 SAMs, downing 3 aircraft.[202]

During early 1972 the North Vietnamese continued the buildup of their forces north of the DMZ, and on the morning of 30 March they launched a triple-pronged invasion of South Vietnam.

The U.S. quickly responded with heavy airstrikes, supported by Wild Weasel and Navy Iron Hand assets, throughout North Vietnam south of the 20th parallel. On 7 April 1972, TAC began the first of a series of combat aircraft redeployments to Southeast Asia under the code name Constant Guard. In the first increment were 12 F-105Gs from the 561st TFS at McConnell AFB, Kansas. They arrived at Korat RTAFB on 12 April and began flying combat missions the same day. By mid-April the air war had moved back into the Hanoi-Haiphong area.

The strike support tactics employed in the first Linebacker air campaign, which commenced in May, differed from those used from 1965 to 1968. During Rolling Thunder chaff was not available to strike forces, but since then chaff bombs and pods that could be carried by fighter aircraft had entered service. Chaff flights, which sowed corridors of chaff to screen the strike aircraft, now preceded the strike force by as much as 20 minutes. Easily identifiable on North Vietnamese radar, the chaff flights were dangerously exposed to SAMs and MiG attacks and required dedicated F-4D/E MiGCAP and Wild Weasel protection. Other MiGCAP and Wild Weasel flights flew with the strike force.

North Vietnamese tactics had evolved since Rolling Thunder, which in turn necessitated changes in Wild Weasel tactics. By 1972 the North Vietnamese had increased their use of coordinated SA-2 site tactics to counter the Shrike and even the Standard ARM. They had also introduced the Fan Song F, which provided the option to use an optical tracking system instead of the E–F-band radar. When optical tracking was employed, only the C-band missile guidance signal was radiated, significantly reducing the capabilities of radar warning systems and Wild Weasel avionics, and precluding the use of antiradiation missiles. Another technique, known as track-on-jam, was used against the ECM-equipped B-52s, if they strayed from their chaff corridors, and also EB-66s on occasion. The SA-2 sites passively tracked the jamming from the bombers in azimuth, and the Fan Song didn't radiate. A missile was then launched in the direction of the jamming (the Fan Song was sometimes switched on during the latter part of the SA-2's flight). In December, during Linebacker II, the North Vietnamese introduced a modified AAA radar to provide ranging data to SA-2 sites for use in conjunction with the track-on-jam technique. This radar (originally given the ELINT designation T-8209 and later dubbed Team Work) could switch between E-band and I-band operation. When operating in I band it was far from the E–F-band frequency range of the North Vietnamese Fan Songs, and its operating characteristics were unrecognizable to RHAW systems and unfamiliar to EWOs in the B-52s and Wild Weasel aircraft.

In the face of these North Vietnamese countermeasures, the Wild Weasels reverted to hunter-killer tactics in combination with radar suppression. The F-105Gs began carrying bombs again in addition to antiradiation missiles, as they had during Rolling Thunder, but this exacted a large performance penalty. They were later paired with F-4E killers, which could not only carry more iron bombs or CBUs than the Weasels, but also carried Sparrow air-to-air missiles for use in the event of a MiG attack. With the advent of the F-4E killers, the F-105Gs returned to carrying only Shrikes and Standard ARMs. Strike force support typically included both F-105G-only suppression and F-105G/F-4E hunter-killer flights. In the suppression role the F-105Gs began to employ preemptive antiradiation missile firings, a tactic that had been employed on only a few occasions during Rolling Thunder. Between April and October alone the Wild Weasels and Navy Iron Hand aircraft fired 574 Shrikes preemptively out of a total of 1,935 launched during the period.[203]

In mid-October the bombing of North Vietnam was once again restricted to the area south of the 20th parallel as the peace negotiations then in progress in Paris seemed to be bearing fruit. Progress in the negotiations proved illusory, however, and in mid-December the North Vietnamese walked out. His patience exhausted, President Richard Nixon then ordered a series of maximum-effort strikes in the Hanoi-Haiphong area. The airstrikes, code-named Linebacker II, began on 18 December 1972.

Wild Weasel F-105Gs and F-4Cs, which had entered combat the previous September, provided suppression support to B-52 strikes, while hunter-killer flights supported tactical air strikes. On the third night of Linebacker II the North Vietnamese began using the T-8209 radar to increase the accuracy of SAMs launched against the B-52s. On the first two nights the track-on-jam technique had been used extensively, but SAMs launched using that mode weren't very accurate. The greatly increased accuracy seen on night three came as a surprise to the B-52 and F-105G EWOs, since it appeared that the SAMs were launched without radar tracking. The presence of an unaccustomed I-band signal was noted but not immediately associated with the SAM launches. Postmission analyses of the I-band signals recorded that night, combined with debriefs of the EWOs, suggested that T-8209 was probably providing range information to SA-2 sites for use in conjunction with passive angle-tracking of jamming emissions. Fortunately the AGM-78B and C missiles covered this frequency range and a new Shrike guidance section, the Mk 36, also operated in I band.[204] This would allow the F-105G Weasels, with their APR-35 panoramic receivers, to recognize and attack the T-8209 during the remainder of Linebacker II. The F-4C Weasels were at extreme risk from this radar, however, since their ER-142 panoramic receivers, identical to those in the pre-Mod 1 F-105F Weasels, had no capability to receive I-band signals.[205]

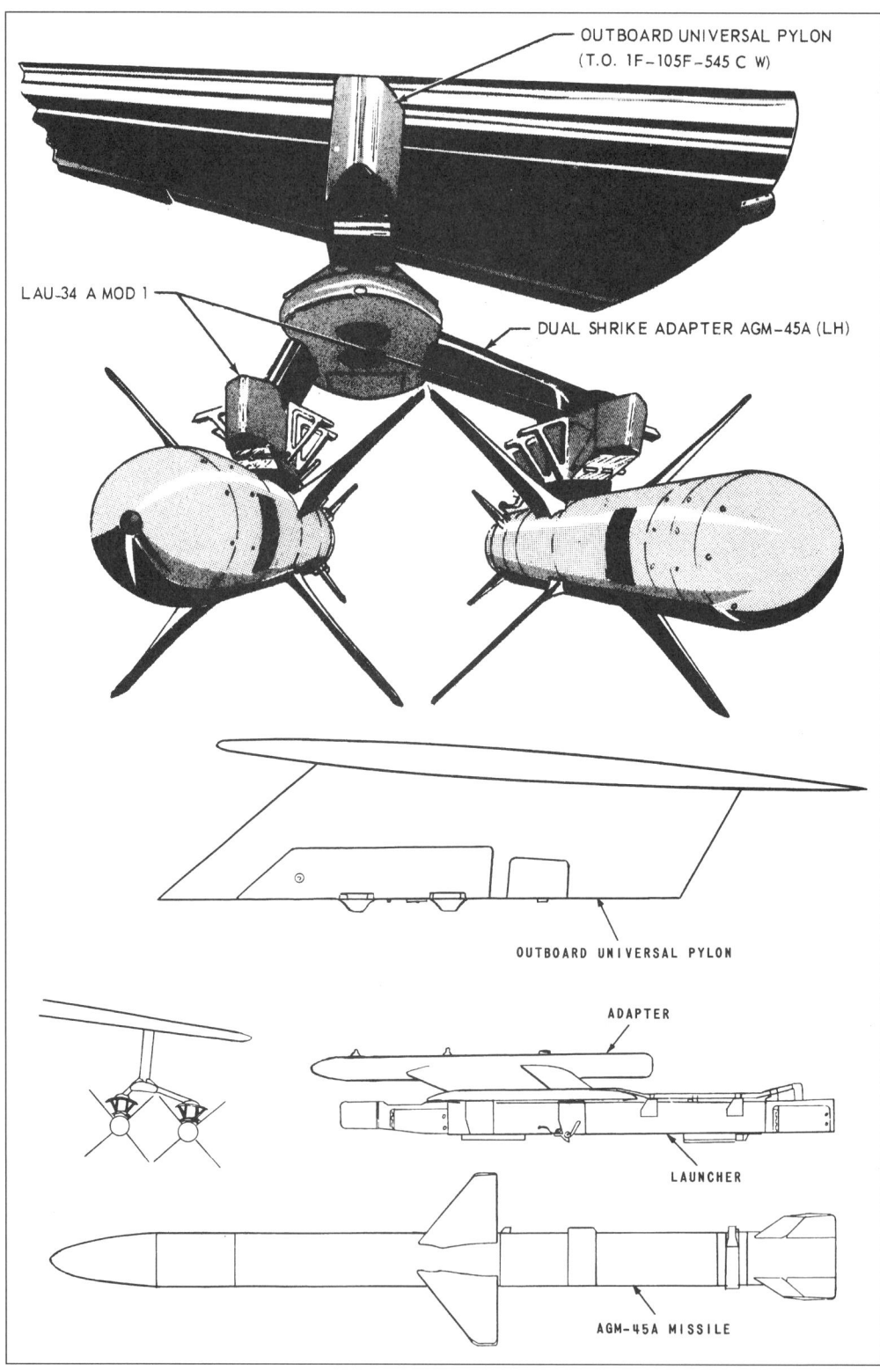

The dual-Shrike launcher was introduced in early 1972 and eventually incorporated in all F-105Gs via TCTO 1F-105F-545. The ADU-315 (right wing) and -316 (left wing) adapters were developed by the Navy at China Lake. The launcher was attached to the standard outboard universal pylon, with the outer missile carried far outboard. The missiles could not be symmetrically located under the pylon since the inner one would have interfered with stores carried on the inboard wing pylon. *(U.S. Air Force)*

Some reports indicate that the dual-Shrike adapter was not used extensively in Southeast Asia, but as this photo of a 17th WWS F-105G-1-RE (62-4434) shows, at least some missions were flown with it. The F-105G in the background is carrying a single Shrike under its outer pylon. *(via Paul Minert via the Mick Roth Collection)*

An F-105G-1-RE (62-4423) takes off loaded with an AGM-45 and AGM-78 under the right wing and a fuel tank under the left wing. Note the absence of the ALQ-105 blisters on the fuselage. The ZB tail codes indicate the aircraft was assigned to the 6010th WWS at Korat. *(Marty Isham Collection)*

Another F-105G-1-RE (63-8301) from the 6010th WWS at Korat. Note the lack of an ALQ-105 blister on the fuselage. The 17th WWS took over the assets of the 6010th WWS on 1 December 1971 as the only remaining F-105 squadron in Southeast Asia. At the end of the war the squadron was reassigned as the 562nd TFS at George AFB, California. *(U.S. Air Force)*

This F-105G-1-RE (63-8333), shown in 6010th WWS markings, was shot down by a North Vietnamese SAM on 17 February 1972 while assigned to the 17th WWS. The crew, Captain J. D. Cutter and Captain K. J. Fraser, were captured and repatriated at the end of the war. *(Don Logan via the Mick Roth Collection)*

An F-105G-1-RE (63-8304) from the 561st TFS at McConnell AFB, Kansas, photographed at Davis-Monthan AFB. Detachment 1 from this squadron deployed to Korat on 7 April 1972 as part of Operation Constant Guard. The detachment was integrated into the operations of the 17th WWS and participated in both Linebacker I and II. During Linebacker II the squadron changed their tail codes to WW. On 5 September 1973 the detachment was relieved from combat duty as part of Operation Coronet Bolo I and was assigned to the 35th TFTW at George AFB, California. *(Gary Meinert via the Mick Roth Collection)*

Sharkmouths became popular on the Weasel aircraft during the war in Southeast Asia. This F-105F-1-RE (63-8339) is from the 17th WWS that was the sole remaining F-105 unit in Southeast Asia after the F-105Ds departed in 1970. The 17th would depart Korat on 29 October 1974, heading for the United States as part of Operation Coronet Exxon. *(Larsen/Remington via the Mick Roth Collection)*

A good look at the upper side of a Thunderchief. The boom refueling receptacle can be seen forward of the windscreen, offset slightly to the left. An AGM-45 Shrike is being carried on each outer wing station. Note the uncamouflaged wingtips—these panels were changed when the ALR-31 antennas were installed, which also involved moving the navigation lights from the wingtips to the forward edge of the wing. *(John Huggins via the Mick Roth Collection)*

Following the Christmas bombing pause, B-52s and F-111s were for the first time used to hit SAM sites, as well as SAM storage areas, in a concentrated effort to decimate North Vietnam's surface-to-air defenses. These attacks, in addition to those of the Wild Weasels and Navy Iron Hand A-6Bs, had reduced North Vietnam's defenses to near impotence by the time Linebacker II ended on 30 December 1972.[206]

The termination of Linebacker II didn't spell the end of the bombing of North Vietnam. After a short New Year's bombing pause, air attacks resumed south of the 20th parallel until mid-January 1973, when the air war over North Vietnam was finally ended. Air operations over Laos and Cambodia continued, however, and Wild Weasel F-105s remained in Thailand until late 1974.

This F-105 (63-8320) was credited with three MiG kills during its career in Southeast Asia. In September 1973 it was wearing three large red stars on a white background. Three different crews accounted for the kills—one by 20-mm gun fire, one with an AIM-9, and the last was downed when the F-105F's pilot jettisoned his centerline MER full of bombs directly into the path of the MiG. *(Hugh R. Muir Collection via the Terry Panopalis Collection)*

Two Very Special Missions[207]

On 10 March 1967, Captain Merlyn Hans Dethlefsen departed Takhli on his 78th combat mission. Over 500 miles away lay the Thai Nguyen steel mill and industrial complex in North Vietnam. Nestled in a valley 40 miles north of Hanoi, and 70 miles from the Chinese border, the heavily defended complex had only recently been approved as a target. Dethlefsen was number three of a four-flight of F-105s sent to disable the air defenses that ringed the target prior to a larger strike force.[208]

The flight leader and Captain Dethlefsen were flying Wild Weasel F-105Fs, while the other two aircraft were D-models. The F-105s rendezvoused with a tanker, refueled, and departed the tanker just five minutes ahead of the main strike force. The flight flew at medium altitude to a point about two miles from the target, then zoomed for altitude and rolled into a steep diving attack on an active SAM site. Dethlefsen and his wingman were nearly a mile behind the first element when they lost sight of the leader in heavy flak. A moment later, the parachute beeper signal on the emergency radio channel confirmed that the lead crew had ejected from their aircraft. Number two reported that his F-105D had also been badly damaged. Dethlefsen took command of the flight,

> We were still ahead of the strike force and they [the strike force] were still vulnerable. We had fuel and missiles, guns and bombs, and the job wasn't done yet. Lincoln lead had seen the target and launched a missile, but it had missed. I decided we would stay. Coming around, I studied the flak pattern. It wasn't a matter of being able to avoid the flak, but of finding the least-intense areas.

Dethlefsen and his backseater, Captain Kevin "Mike" Gilroy, had located the approximate position of the SAM site on the first pass. As he maneuvered the aircraft to line up with the target, Dethlefsen spotted two MiGs closing fast from the rear. He fired an AGM-45

An F-105G-1-RE (62-4432) and F-105D-15-RE (61-0073) prepare to take off at Nellis AFB in 1973. Note the large TAC badges on the vertical stabilizers; the G-model also has a 57th FWW badge on the forward fuselage. *(Mick Roth)*

Shrike missile at the SAM site and veered sharply away as one of the MiGs launched a missile toward the F-105.

The Thunderchief was no match for the maneuverable MiG in a dogfight. Therefore, it was standard procedure for F-105 pilots under attack by MiGs to jettison their ordnance, engage the afterburner, and head for the treetops where the F-105 could easily outacceler-

The Sacramento Air Materiel Area (SMAMA) at McClellan AFB was the depot for all F-105s after 1966. This 57th FWW F-105G-1-RE (62-4438) taxis past work hangers filled with other F-105s. Note the extended refueling probe. *(Mick Roth)*

ate the interceptors. However, as Dethlefsen positioned for another pass, he saw two more MiGs and evaded them with a tight left break. He elected to retain his ordnance, but 57-mm AAA had hit Dethlefsen's F-105. Fortunately, shrapnel through the bottom of the fuselage and the left wingtip had not damaged any vital systems. The two-man crew again turned their attention to the SAM site, and Dethlefsen remembers that the main force was already leaving the industrial complex.

"I could hear the strike force withdrawing. I had permission to stay there after they left. That steel mill with the related industry was too big a target—too big to knock out with one strike. I knew those fighter-bombers would be back tomorrow. Same route, right over this area. My aircraft was working well enough to be effective. With the weather the way it was that day I knew we would never have a better chance. So I made up my mind to stay until I got that SAM site, or they got me."

Maneuvering around the flak, Dethlefsen spotted another SAM site and fired an AGM-45. Smoke and dust from the main strike on the complex began to drift over the defensive positions as the Dethlefsen and Gilroy attempted to reacquire the original SAM site, so Dethlefsen eased the F-105 lower for a better look. During the last pass, Dethlefsen dropped his bombs squarely on the site and followed with 20-mm cannon fire. Major Kenneth Bell, his wingman, had stuck to him like glue, but Bell's F-105 had been hit by both AAA and a MiG.

The two battle-weary F-105s sped toward the tanker and Takhli. "All I did was the job I was sent to do," Dethlefsen said. "It had been quite a while since we had been able to go into the Hanoi area. So while the weather held we were able to do some pretty good work. It was a case of doing my job to the best of my ability. I think that is what we mean when we call ourselves professional airmen in the Air Force."

Six weeks after Captain Dethlefsen flew his mission over Thai Nguyen, another Weasel team took off from Takhli. Major Leo K. Thorsness Captain Harold E. Johnson were nearing the 100-mission mark that would bring their combat tour to an end. MiGs had chased Thorsness and Johnson through the skies of North Vietnam, and the experienced pair had evaded 53 SAMs.[209]

Describing Weasel missions, Thorsness said, "In essence, we would go in high enough to let somebody shoot at us and low enough to go down and get them; then went in and got them." The Weasels would be the first flight on target, preceding the main attack and remaining after the strike force had departed.

On 19 April 1967, Thorsness led four F-105s from a tanker in southern Laos toward the Xuan Mai army barracks and storage supply area 30 miles to the southwest of Hanoi. Xuan Mai lay on the edge of the Red River delta, and hopefully the defenses would not be as lethal as those ringing downtown Hanoi.

Thorsness sent number three and four to the north of Xuan Mai as he and number two headed to the south. Now the enemy gunners would be forced to divide their attention between the separated elements. Thorsness maneuvered toward a strong SAM signal and fired a Shrike. The site was seven miles distant and obscured by haze, so Thorsness and Johnson never saw the missile hit.

Thorsness picked up a second SAM site visually and attacked it with a CBU-24 cluster bomb. The two F-105s accelerated toward the treetops, but AAA had hit the engine of Tom Madison's F-105. Moments later the rescue beeper on guard (emergency) radio channel signaled that Madison and Tom Sterling had ejected. In the midst of all this, Thorsness managed to fire another Shrike at a third SAM site.

To the north, the other two F-105s had survived an encounter with MiGs, although number three's afterburner would not light, and without the added thrust the flight could not sustain the supersonic speed to outrun their attackers. The pair headed back for Takhli. Thorsness was now the only fighter-bomber in the Xuan Mai area.

Thorsness circled the descending parachutes of Madison and Sterling while Johnson relayed information to "Crown," the rescue control aircraft. Suddenly Johnson spotted a

This F-105G-1-RE (63-8305), assigned to the 562nd TFS, was photographed at George AFB in July 1978. *(Mick Roth)*

Two F-105G-1-REs (63-8266 in the background, and 62-4428) from the 35th TFW at George AFB. The 35th would later give up the F-105Gs to the Georgia ANG, although by that time the aircraft wore WW tailcodes. *(Bill Malerba via the Mick Roth Collection)*

The WW tail codes simply would not go away. Beginning in the fall of 1979, the GA tail codes were replaced with first white, then black, WW codes. This F-105G-1-RE (63-8307) was assigned to the 562nd TFS at George AFB. With the end of air operations in Southeast Asia, all Wild Weasel F-105Gs and F-4Cs except for a few assigned to the 66th FWS at Nellis were consolidated at George. In September 1975, Nellis gave up its last Weasels to the 35th TFW at George, where they stayed for the remainder of their active Air Force service. *(Mick Roth)*

MiG off their left wing, and Thorsness recalls, "I wasn't sure whether or not he was going to attack the parachutes. So I said, 'Why not?' and took off after him. I was a little high, dropped down to 1,000 feet and headed north behind him. I was driving right up his tailpipe at 550 knots. At about 3,000 feet I opened up on him with the 20-mm but completely missed him. We attacked again, and I was pulling and holding the trigger when

Sharkmouths began to reappear on F-105Gs at George AFB in May 1980, shortly before they were retired from active Air Force service. The sharkmouth shown here is decidedly different from those used in Southeast Asia. *(Mick Roth)*

Harry got my attention with the MiGs behind us. If I had hit that MiG good we would have swallowed some of the explosion (debris). But we got him."

Low on fuel, Thorsness headed south toward a tanker, following the progress of the rescue forces on his radio. The prop-driven A-1 "Sandys" that would direct the on-scene effort and the rescue helicopters that would attempt the pick-up were already headed toward the downed Weasel crew. With full tanks but with only 500 rounds of ammunition, Leo left the tanker and flew north again. While briefing the Sandy pilots on the defenses around Xuan Mai, he spotted three MiGs ahead. "One of the MiGs flew right into my gunsight at about 2,000 feet and pieces started failing off the (enemy) aircraft. They hadn't seen us, but they did now."

Johnson warned that four MiGs were closing from the rear and Thorsness dove for the deck, eluding his pursuers as the F-105 raced through the mountain passes in full afterburner. Now the MiGs turned back toward the slow-moving Sandys, and Thorsness radioed, "Okay, Sandy One. Just keep that machine of yours turning and they can't get you." Low on fuel again and without ammunition, Leo turned toward the MiGs with one idea in mind: "To try to get them on me." Soon, another flight of F-105s arrived, and now the MiGs were on the defensive.

But Thorsness was far from satisfied as he flew toward the tanker for the third time. A Sandy had been shot down, and the rescue effort for Madison and Sterling had been called off. Both were later captured. Now, another F-105 pilot called Thorsness: "Leo, I'm not with the rest of the flight, and I don't know where I am. I've only got 800 pounds [of fuel]. What should I do?"

What Thorsness did was send the tanker he was supposed to refuel from north toward the lost pilot. It was another courageous act by a tanker crew over Southeast Asia. Thankfully, the rendezvous was successful and the pilot plugged into the tanker before the engine flamed out from fuel starvation. But now Thorsness was critically low on fuel as he continued south toward the nearest recovery base at Udorn, Thailand.

> I knew if we could get to the Mekong River—the Fence—we could coast across. With 70 miles to go, I pulled the power back to idle and we just glided in. We were indicating 'empty' when the runway came up just in front of us, and we landed a little long. As we climbed out of the cockpit, Harry said something quaint like, 'That's a full day's work!'

For their actions on these missions, Captain Merlyn Hans Dethlefsen and Major Leo K. Thorsness were both awarded the Medal of Honor.

The End

With the end of air operations in Southeast Asia, all Wild Weasel F-105Gs but for a few assigned to the 66th FWS at Nellis AFB were consolidated at George AFB, California. In September 1975 Nellis gave up its last Weasels to the 35th TFW at George, where they stayed for the remainder of their active Air Force service.

The last F-105G Weasel avionics improvements began before the end of air operations in Southeast Asia with the QRC-535 and -555 upgrades of the APR-36 and -37. QRC-535 improved the APR-37's missile guidance processing capabilities and extended its C-band coverage into D band. QRC-555 replaced the APR-36's four video amplifier-detectors with units that provided E–J-band coverage, enhanced the pulse analyzer's threat detection logic, and provided more explicit threat light placards. The later replacement of the APR-36/QRC-555 pulse analyzer with the QRC-72-06 digital processor (first tested in the F-105G in late 1973) upgraded the system to the ALR-46 configuration. The first operational, fully digital radar warning receiver, the ALR-46 provided C–J-band frequency coverage and included both tracking and missile guidance radar reception and identification capabilities. Where the

Beginning in 1979 the F-105Gs with the greatest remaining airframe life were transferred to the Georgia ANG—the rest were sent to MASDC at Davis-Monthan AFB in Tucson, Arizona. The Southeast Asia-style sharkmouth made a return after the F-105Gs (63-8265) were assigned to the Georgia ANG. *(Fred Harl via the Mick Roth Collection)*

As the aircraft were repainted they received a wrap-around camouflage, still with the sharkmouth. Here an F-105G-1-RE (62-4439) shows the small serial number the Georgia ANG used, very reminiscent of the style used in 1967 when the aircraft were first camouflaged. The Georgia ANG performed a number of modifications to their F-105Gs. The az-el/homing antennas were removed, as were the strike camera and all components of the ALR-31 system. The original F-105F wing tips were restored since the ALR-31 antennas were removed. Another modification gave the aircraft the capability to captive-carry an AGM-78 on the centerline station. *(Mick Roth Collection)*

Most Georgia ANG aircraft carried the ANG badge on the vertical stabilizer, but some did not at various times. Here an F-105G-1-RE (62-4423) in wraparound camouflage is parked next to an aircraft still in its Southeast Asia-style paint with gray undersides. *(Kirt Minert via the Mick Roth Collection)*

Not all the aircraft had been repainted before they were retired from the Georgia ANG. When PEACH 91 landed at Dobbins AFB on 25 May 1983, it marked the end of the F-105's career. The aircraft carried special "Last ANG F-105 Flight—1964–1983" markings on the fuselage under the cockpits. The area of the fuselage under the wings (where it was gray) was signed by all the Georgia ANG personnel. Interestingly, the aircraft is an F-105F-1-RE (63-8299), not a Weasel. *(Terry Panopalis Collection)*

previous systems had displayed dotted, dashed, and solid radial strobes on a cathode ray tube azimuth indicator to generically identify and indicate the approximate bearing and signal strength of threats, the ALR-46 displayed unique alphanumeric symbols. With the ALR-46's expanded capabilities, the ALR-31 was no longer required and was later removed from the F-105G.

Even with the ALR-46 upgrade, the F-105G's Weasel suite was as dated by the mid-1970s as was the F-4C Weasel's during Linebacker II. Threat frequencies had proliferated and the signal densities generated by the new threat radars had increased so much that the F-105G was no longer capable of effectively operating against them. The next-generation Wild Weasel, the F-4G, entered service in 1978 equipped with the fully digital APR-38 radar warning and attack system. The APR-38 provided capabilities far beyond those of the F-105G's systems, and the introduction of the F-4G brought to an end the active Air Force career of the Weasel Thunderchief.

Beginning in 1979 the F-105Gs with the greatest remaining airframe life were transferred to the Georgia ANG, and the remainder were sent to the boneyard at Davis-Monthan AFB in Tucson, Arizona. The F-105G's service with the Georgia ANG was not long, however. Even the best of the F-105Gs were tired and worn, and spare parts were in short supply. Despite a series of modifications to extend the service lives of the aircraft, the end came in 1983 when the last of the F-105Gs were retired.

Notes

141. *DRV SAM Problem,* USAF Combat Arena Background Paper, undated, p 1.
142. Nalty, Bernard C., *Tactics and Techniques of Electronic Warfare: Electronic Countermeasures in the Air War Against North Vietnam,* Office of Air Force History, 16 August 1977, p 32.
143. *DRV SAM Problem,* pp 3–4.
144. Nix, Lieutenant Colonel Cowan G., *Wild Weasel I: Response to a Challenge,* USAF Southeast Asia Monograph Series, Air War College, Maxwell AFB, Alabama, June 1977, pp 30–31.
145. The code name of the F-100F project was later changed to Wild Weasel I when the Air Force began developing other aircraft to perform the specialized anti-SAM mission.
146. *Interim Report on APGC Project 669AY3, AN/APS-107 (XA-1) Operational Evaluation,* APGC, Eglin AFB, Florida, 27 December 1965, p 3.
147. *TAC-AFSC Anti-SAM Test Program,* Briefing Paper, 6 October 1965.
148. CSAF Message AFSME 86001, Subject: Radar Homing and Warning (RHAW) for F-105D/F Aircraft, 15 October 1965.
149. *SMAMA Participation in Southeast Asia Build-Up,* SMAMA Historical Study No. 61 (Part II), SMAMA History Office, McClellan AFB, California, September 1967, pp 171–172.
150. *Ibid.*
151. *Ibid.*
152. *Anti-SAM Task Force November Review,* 3 November 1965
153. TAWC Message FT 08232, Subject: 13 Anti-SAM Tests, 1 November 1965.
154. SMAMA Message SMNE 24727, 3 November 1965; *Project Directive for Project 0435Y, APS-107B/ALQ-51/QRC-160-1 Evaluation,* APGC, Eglin AFB, Florida, 8 December 1965, p 3.
155. *Operational Test & Evaluation of the AN/DPN-61 Radar Homing Set (Modified),* APGC, Eglin AFB, Florida, September 1965, pp 2–3.
156. *Wild Weasel IA Final Report,* APGC TR-66-3, APGC, Eglin AFB, Florida, January 1966, pp 3, 7.
157. *Ibid.,* pp 3, 16, 33–35.
158. *Ibid.,* pp ii, 7–10, 16, 29, 31; C.K. Bullock audiocassette tape recorded for Larry Davis in 1983.
159. *Wild Weasel IA Final Report,* p 10.
160. TO 12P3-4-24-2, *ATI IR-133C Intercept Receiver System Maintenance with Illustrated Parts Breakdown,* 30 June 1966.
161. Joint Electronic Type Designation Technical Data Card, AN/APR-26(V), 12 December 1965; *The Vector in a SAM Environment,* 355th TFW, 7 May 1966. pp 5, 12, 15.
162. Applied Technology, Inc. Letter 401-66-625/10291 to Colonel J.T. Johnson, SMAMA, 17 March 1966, Subject: F-105F Wild Weasel Installation.

163. *Ibid.*
164. *Ibid.*
165. *SMAMA Participation in Southeast Asia Build-Up,* pp 174–175.
166. Enough video cable was available for the IR 133 installation, which required far less than the APR-25.
167. ATI Letter 401-66-625/10291.
168. *SMAMA Participation in Southeast Asia Build-Up,* pp 174–175.
169. Wild Weasel III Rework Team Report, SMAMA, McClellan AFB, California, 15 March 1966.
170. Wild Weasel III Rework Team report; ATI Letter 401-66-625/10291.
171. Minutes of RHAW and Wild Weasel Plans and Guidance Meeting, 29–30 March 1966, SMAMA, McClellan AFB, California, Attachment 3.
172. CSAF Message AFRDQR 91635, Subject: Added Capability for F-105F Wild Weasel III (Roman) Aircraft, 15 February 1966; *RHAW Testing for Tactical Aircraft From 11 October 1965 through 29 March 1966,* Working Paper, TAWC/APGC, Eglin AFB, Florida, pp 49–50, 56, 58; TO 12P3-4-24-2.
173. In operational use the az-el system proved to be somewhat difficult to use and troublesome to maintain, but it was better than nothing. Az-el variants were used in all subsequent Wild Weasel F-105F/Gs, but problems apparently persisted, since a number of photos exist of Weasel Fs in Southeast Asia with cover plates instead of homing antennas. The APR-35 homing antennas were eventually removed from all Georgia ANG F-105Gs.
174. *Project SEE-SAMS, Phase IV, APGC Project 0398T,* APGC, Eglin AFB, Florida, July 1966; *RHAW Testing for Tactical Aircraft,* pp 9–10, 56, 58.
175. *Project SEE-SAMS.*
176. *Ibid.*
177. *RHAW Testing for Tactical Aircraft,* p 16; Monthly Anti-SAM Test Program Status Reports, APGC, Eglin AFB, Florida, 31 March and 30 April; APGC Message PGTR 10117, Subject: APR-26 Deficiencies, 5 May 1966.
178. Paul Chesley audiocassette tape recorded for Larry Davis in 1983.
179. E-mail from Colonel Edward T. Rock, USAF (Ret.), October 1999.
180. *Ibid.*
181. *F-105 Combat Tactics,* 388th TFW, Chapter V (dated 20 May 1967) and Attachment B (Undated).
182. TAC Message DO 52775, Subject: Wild Weasel III and IVC Training, 10 September 1966.
183. AFLC F-105B/D/F D-20 Logistics Report, SMAMA, McClellan AFB, California, 31 August 1966, p VII–VI.; *Wild Weasel IV-C Final Report,* TAC TR-65-85C, TAWC, Eglin AFB, Florida, December 1967, pp 3, A-3-A-5, B-3.
184. CSAF Message AFRDQRT 87037, Subject: Wild Weasel III/Mod Request 1778, 11 August 1966; AFLC F-105B/D/F D-20 Logistics Reports, 31 December 1966 and 30 June 1967.
185. Danforth, Colonel Gordon E., *Iron Hand/Wild Weasel,* Corona Harvest Special Report, Aerospace Studies Institute, Air University, Maxwell AFB, Alabama, January 1970, pp 10–11.
186. *Ibid.;* Paul Chesley audiocassette tape.
187. *F-105 Combat Tactics,* Chapter V, pp 10–11.
188. AFLC F-105B/D/F D-20 Logistics Report, 31 May 1967, p VIII-ii. It is possible that the EF-105F designation may have been reserved for the Combat Martin "Special ECM" aircraft. The 1968 volumes of the 388th TFW and 44th TFS histories refer to the Wild Weasel aircraft as F-105Fs and the Combat Martin aircraft as EF-105Fs. Information on whether the designation was informal or based on a request for redesignation is unavailable in material declassified to date.
189. AFLC F-105B/D/F D-20 Logistics Reports, 30 September 1967 and 31 July 1968.
190. *SEAOR Status Report,* SEAOR 8 (FY66), Department of the Air Force, Directorate of Operational Requirements and Development Plans, Multiple dates from 6 May 1966 through December 1967.
191. *SEAOR Status Report,* SEAOR 88 V Mod (FY67), December 1967; *History of the United States Air Force Tactical Fighter Weapons Center,* 1 January through 30 June 1967, p 46.
192. *SEAOR Status Report,* SEAOR 8 (FY66), December 1967.
193. *SEAOR Status Report,* SEAOR 2 (FY66), 18 August 1967; *History of the United States Air Force Tactical Fighter Weapons Center,* p 47.
194. *History of the United States Air Force Tactical Fighter Weapons Center,* p 47; Jack Donovan audiocassette tape recorded for Larry Davis on January 29, 1983.
195. 44th TFS History, 1 October to 31 October 1968. Four F-105Fs (63-8281, 63-8285, 63-8327, and 63-8347) are known to have been modified for BASS I/II testing.
196. *F-105 Combat Tactics,* Attachment B, pp 3–4.
197. AFLC F-105B/D/F D-20 Logistics Report, 30 September 1967.
198. Danforth, *Iron Hand/Wild Weasel,* p 13.

199. AFLC F-105B/D/F D-20 Logistics Report, 31 December 1968.
200. Hewitt, Major William A., *Planting the Seeds of SEAD: The Wild Weasel in Vietnam*, Air University Press, Maxwell AFB, Alabama, June 1993, p. 18.
201. *SEAOR Status Report*, SEAOR 8 (FY 66), 4 and 18 August 1967; TO 1F-105B-34-1-1S-105, 30 March 1972.
202. Nalty, *Tactics and Techniques of Electronic Warfare*, pp 6–7, Zaloga, Stephen J., *Soviet Air Defense Missiles*, Jane's Information Group Inc., Alexandria, Virginia, 1989, pp 60–62.
203. Hewitt, *Planting the Seeds of SEAD*, p 19; Nalty, *Tactics and Techniques of Electronic Warfare*, pp 96–98.
204. The Mk 36 guidance section was used in the AGM-45A-6 (and later AGM-45B) tactical Shrike. The dash number assigned to a tactical Shrike missile (e.g., AGM-45A-6) was determined by the specific guidance section used. AGM-45As used a Mk 39 or Mk 53 single-thrust rocket motor; AGM-45Bs (introduced around 1973) used the Mk 78 dual thrust rocket motor.
205. Eschmann, Karl J., *Linebacker: The Untold Story of the Air Raids Over North Vietnam*, Ivy Books, New York, November 1989, pp 136–137, 149–150.
206. Eschmann, *Linebacker*, pp 150–151, 163–192.
207. These are very condensed versions of the two F-105 Medal of Honor missions. Fuller versions have been printed in several magazines and books, and the full text is available in the Monograph referenced in the next citation.
208. USAF Southeast Asia Monograph Series Vol. VII Monograph 9, Air Force Heroes in Vietnam, pp 25–29.
209. *Ibid.*, pp 29–31.

CHAPTER 6

The Details—Construction and Systems

The F-105 was an all-metal, high-performance, swept-wing fighter-bomber designed and manufactured by the Republic Aircraft Corporation in Farmingdale, New York. Republic later became the Republic Aircraft division of Fairchild Hiller Corporation.

Cockpit

The F-105 had a crew of one (B and D) or two (F and G) in separate, pressurized, and air-conditioned cockpits. The electrically operated canopies opened 75° on the F/G and 42° on the single-seat aircraft to provide an equivalent unobstructed area for ingress and egress. The canopies were made up of two plastic panels with an intervening airspace filled with dry air to prevent canopy fog. All F-105s included provisions for a two-piece thermal radiation shield that could be secured to all transparent areas of the canopy. This would, in theory, shield the pilot from the thermal effects of the nuclear weapon he had just released. The aft instrument panel in the F-model was mounted 3 inches higher and 6 inches closer than the front panel, and the top of the panel was also tilted by 9° towards the pilot in an attempt to make it easier to read. This combination of dimensional and angular displacements gave the rear-seat pilot a very disconcerting optical illusion that climbs were steeper than usual, or that descents were shallower.

There were some limitations on what actions the rear crewmember could perform in the F-105F. For instance, he could not trim the aircraft for takeoff, extend the ram air turbine, test the air data computer, change the afterburner range of operation, start the engine, initiate or terminate an air refueling operation, initiate deicing of the engine or pitot tubes, deploy the drag chute, control the IFF/SIF equipment, control the AGM-12, or perform any task that required the use of an auxiliary or override system (since these were only installed in the front cockpit). There were, however, six "take-over" switches installed in each crew compartment to permit either crewmember to independently assume command of certain functions. Each switch was a push-to-operate type with integral lights to provide a visual indication of which crew station is in command of a particular function. The switch

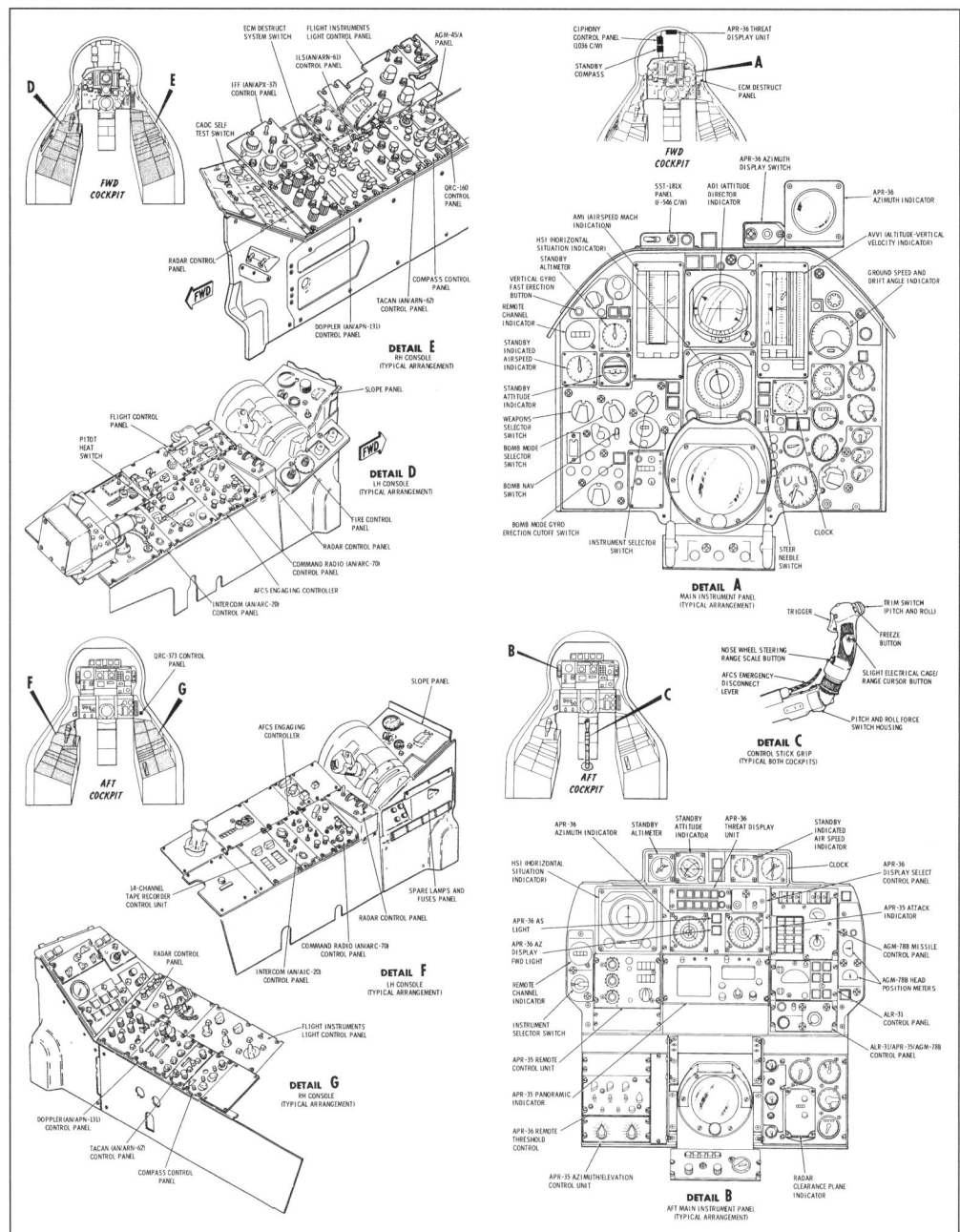

Shown here is a 1970-vintage Wild Weasel F-105G cockpit. The front cockpit of the two-seat F-105 was essentially identical to the single-seat F-105D. In the F-105F trainer, the aft instrument panel was essentially identical to the forward panel. Note the almost complete lack of flight instruments in the Weasel aft cockpit. *(U.S. Air Force)*

in the cockpit in control remained in a depressed position, while the other switch popped out to the normal position.[210]

Each crewmember had an upward firing, catapult-type ejection seat. The seat incorporated an armored headrest, provisions for electrical adjustment of seat height, and adjustable armrests. An explosive-type seat catapult, attached to the rear of the seat, supplied the propelling force to eject the seat. On F-105D-5-RE and later aircraft, provisions were fitted to replace the explosive catapult with a rocket-type seat, and this was done in at least some aircraft modified under TCTO 1F-105-945. With the catapult seat, the recom-

Switch	Command of
FLIGHT	Automatic flight control system (AFCS, instrument landing system (ILS)), instrument mode select switch, and steering needle switch.
NAV	Doppler navigation system and compass system
TACAN	TACAN ground reference navigation system
COMM	UHF command radio set
RADAR	R-14A radar set
WEAPONS	Weapon select switch, bomb mode select switch, bomb/nav switch, bomb mode erection cutoff switch, range wind potentiometer, range wind manual/automatic select switch, six weapon stations select switches, pylon sequence select switch, master armament switch, and freeze/fire switch.[211]

mended ejection altitude was 2,000 feet. The rocket-powered seat provided meaningful ejection at zero altitude as long as there was at least 100 knots airspeed. In the F-105F, ejection was individually controlled from each cockpit and there was no built-in controlled sequence, although it was recommended that the aft crewmember eject first.[212]

Canopy jettison was initiated by pulling upward on handles on either side of the seat bottom. When actuated, both armrests were moved to the raised position, the shoulder harness and leg braces were pulled tight, the canopy locks were opened, and the canopy opened and separated from the aircraft. A second handle was then available to initiate ejection. When the seat was ejected from the aircraft, the emergency oxygen and IFF signal were actuated, the chaff dispenser box was opened (if installed), and a 1e-second safety belt timer initiated. At the end of this time, the seat belt and leg braces were released, the seat/man separator strap assembly displaced the pilot from the seat, and the pilot's parachute set to open. The parachute opened 1 second later if below 14,000 feet, or when the pilot reached 14,000 feet if above that altitude. An optional zero delay lanyard could override the timer.

Engine bleed air was used by a cockpit air-conditioning and pressurization system. Air was ducted to the windshield (defrost), two ducts located beside the seat, and two foot warmers (hollow perforated tubes on which the rudder pedals rode). Cockpit pressurization was automatically maintained, and a temperature control system allowed the pilot to preset the desired temperature. One 10-liter liquid oxygen container and converter supplied gaseous oxygen to the crew in all F-105s. It is interesting to note that the oxygen supply was not increased in the F-105F. (Republic pointed out that the "supply is sufficient to accomplish the equivalent of the 'Project Flying Fish' operation with two crew members—i.e., nonstop ferry flights from California to Okinawa with five air refuelings en route, programmed for cruise at altitudes of 25,000 and 30,000 feet for 14.95 hours."[213])

Fuselage

The fuselage was a semimonocoque structure of 75ST aluminum alloy with a "wasp-waist" at the wing juncture in accordance of the area rule principle. Limited use was made of steel, titanium, and magnesium alloys in areas subjected to high stress or high temperatures. The fuselage consisted of four main parts: the nose, center section, aft section, and the removable rear fuselage. The nose was built up from approximately circular formers, with the skin containing many dielectric and access panels. The pressurized cockpit was above and slightly aft of the nose gear well. Partial frames and angle-section stringers made up the center section, which had an internal weapons bay that was 190 inches long, 32 inches wide, and 32 inches deep. Twin upward-hinged weapons bay doors could be opened at speeds up to Mach 2. This section contained the wing attach points, with the heaviest loads

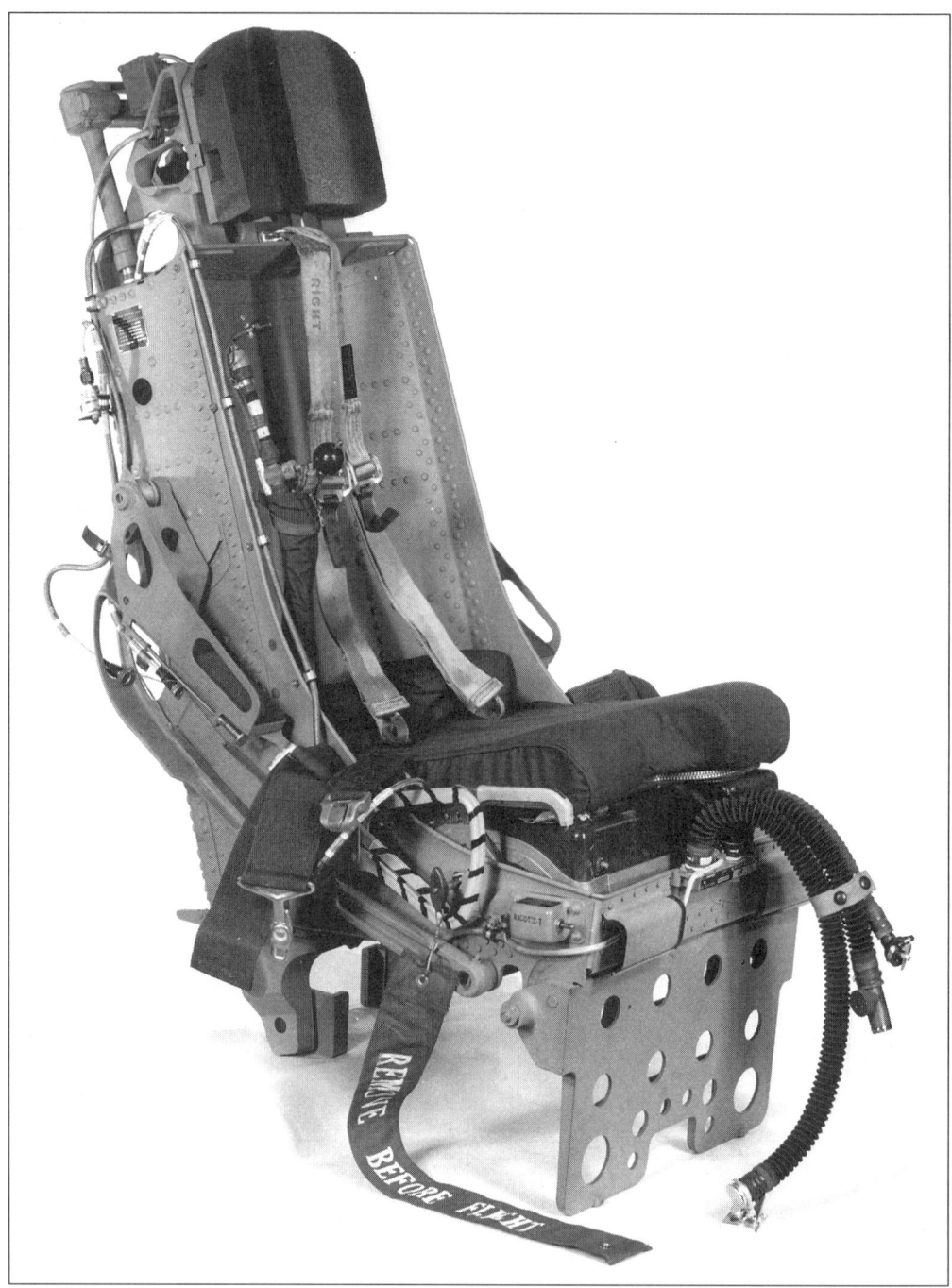

The F-105 ejection seat. The striped handle along the seat cushion initiated the ejection sequence. There were several variations to this seat in later service, each improving areas shown to be deficient in operation. This seat does not have the chaff dispenser that would be added above the pilot's left shoulder. *(Mike Machat Collection via Tony Landis)*

being absorbed by the main and forward H-frames. The rear fuselage was detachable at a point just forward of the wing trailing edge (at station 633) to provide access to the engine. The rear fuselage was attached with four tension fittings and contained the aft fuel cells and ancillary equipment. All aircraft except the second YF-105A (54-0099) had hydraulically operated four-petal speed brakes, made of titanium and stainless steel, that formed the last 3 feet of the fuselage around the engine exhaust. These speed brakes opened 9° when the afterburner was ignited to produce a larger ejector nozzle for the engine. With the

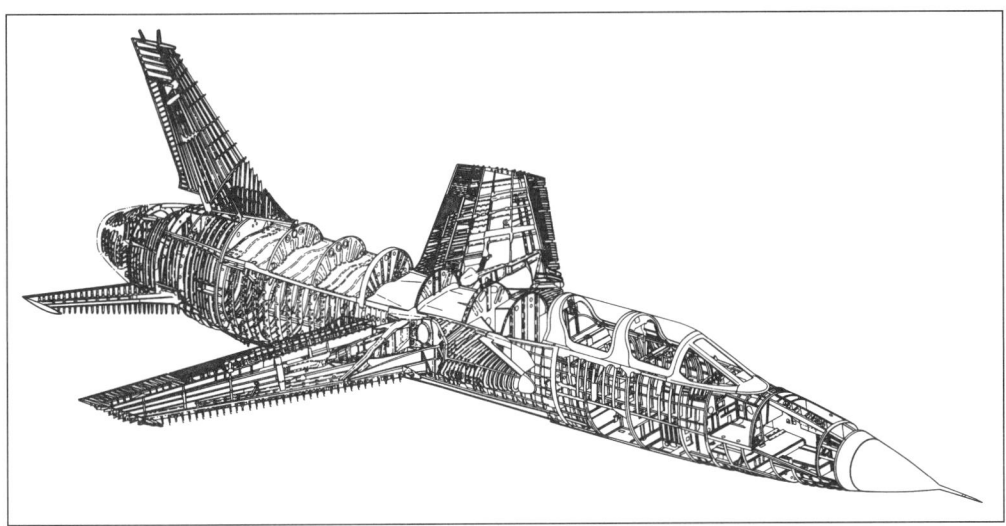

The construction of the F-105 was largely conventional, with the only major advancement being the use of machine-milled skins. This was an expensive process, but lightened the aircraft considerably, while maintaining high structural strength. Machine milled skin was used on all major panels of the fuselage, wings, and empennage. *(U.S. Air Force)*

landing gear extended, only the horizontal petals could be extended, while all four petals could be fully opened with the gear retracted. The lower petal normally drooped somewhat as hydraulic pressure bled down while the aircraft was on the ground.

At its design gross weight of 31,392 pounds, and without external stores, the limit load factors were +8.67/–4.0-g for subsonic maneuvering, and +7.33/–3.0-g for supersonic maneuvering. The F-105 departed from the then industry standard practice of constant thickness skins. Instead, Republic pioneered the concept of machine milling each skin panel to the exact thickness needed for the strength required. This was an expensive process but lightened the aircraft considerably while maintaining high structural strength. Machine-milled skin was used on all major panels of the fuselage, wings, and empennage.

Wings

The wing was of the full-cantilever type, utilizing stress-skin construction, and was composed of two principle spars running from root to tip, with a lighter spanwise member between them. Both front and rear spars were machined from 75ST aluminum alloy. Most of the wing structure was covered by machined skin, pressed to the required contour and attached by bolts and flush rivets. The wing section was an NACA 65A-005.5 at the root and an NACA 65A-003.7 at the wingtip. The last two figures indicated the thickness ratios: 5.5 percent and 3.7 percent. The wing had an anhedral of 3° 3′, with zero incidence, and a leading edge sweep of 45°. A total of 385 square feet of effective wing area was provided, resulting in a nominal takeoff wing loading of 120 pounds per square foot for early F-105Bs. Conventional ailerons, of aluminum alloy construction, were used only at subsonic speeds. Primary roll control was by five sections of hydraulically actuated aluminum alloy spoilers on the upper surface of each wing. All control surfaces were actuated by fully powered, irreversible tandem jacks. No trim tabs or deicing systems were fitted. The glide ratio of the F-105D, distance to altitude, worked out to about 7.75:1, meaning that from 40,000 feet, the Thunderchief could glide for over 58 miles.

Lift-producing devices included aluminum alloy Fowler-type trailing-edge flaps, and full-span aluminum alloy leading-edge flaps. The leading-edge flaps were of conical camber, with the camber increasing progressively from root to tip. The leading-edge flaps were used to

The rear fuselage of an F-105D-10-RE (60-5375) shows the arresting hook integrated into the ventral fin, the large air scoop added to the side of the fuselage, a partially open drag chute door, and the configuration of the petal speed brakes. This aircraft also has the small fairing on top of the fuselage that indicates it has been equipped with the extra hydraulic lines used by the aircraft recovery system. *(C.E. Anderson)*

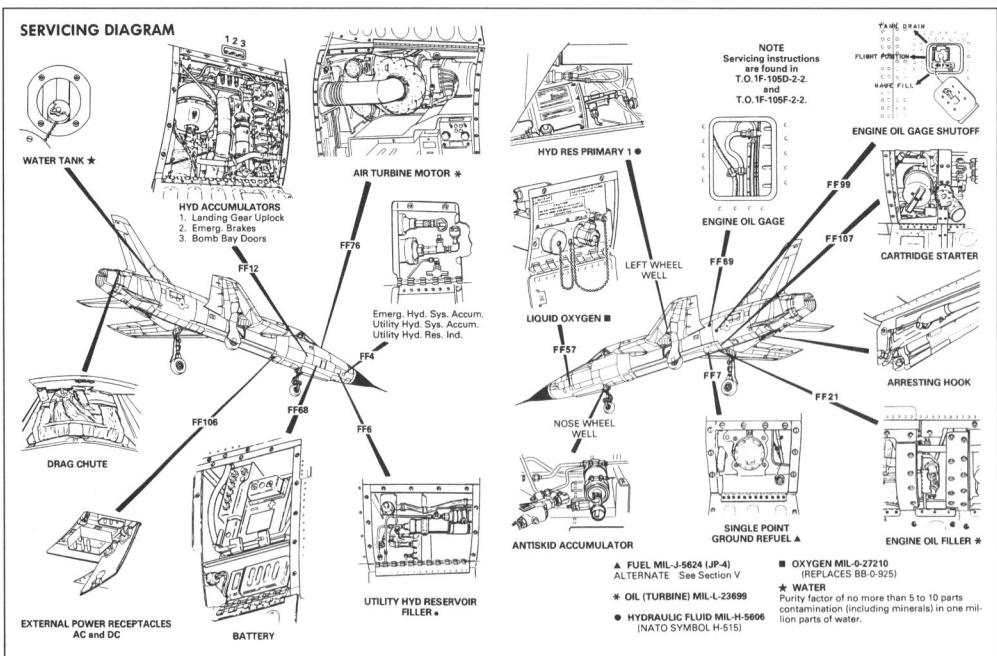

The servicing diagram for a late F-105D shows the various access panels and systems that need to be checked between every mission. Note the water tank on the upper-right aft fuselage for the J75's water injection system, and the provisions for the cartridge starter that allowed the engine to be started without ground power. *(U.S. Air Force)*

increase lift during landing and takeoff, reduce drag during cruise, and improve control during maneuvering flight. Each wing used three actuators, and the leading-edge flaps moved through a maximum arc of 20° and required approximately 10 seconds to fully extend or retract. The trailing-edge flaps were mounted between the wing root and aileron and had a maximum downward travel of 34.5°. Two mechanically interconnected electric motors powered by the dc bus actuated the flaps. An airspeed-sensing switch automatically retracted the flaps if the airspeed exceeded 275 KIAS. On F-105D-30-RE and later aircraft (and those modified by TCTO 1F-105-670), a fail-safe circuit compared left and right flap travel and interrupted power to the flap motors if flap movement diverged by more than 2° during the first third of extension or 7° during the last third. The failure of one flap-actuating motor would cause the flaps to extend slower than usual and might have prevented them from extending completely. Selecting the LANDING setting on the flap handle while in flight would automatically extend the flaps when airspeed went below 260 KIAS. On aircraft modified by TCTO 1F-105-1049 (Aircraft Recovery Mode), a switch in the cockpit allowed the pilot to differentially deploy the flaps for roll control in the event the stabilator was damaged.

All F-105s except the first two prototypes had swept-forward air intake ducts in the wing root leading-edges. These incorporated a variable area inlet (VAI) system that attempted to match the airflow to the engine's requirements over a broad range of speeds and altitudes. VAI used a set of movable, contoured plugs whose adjustment changed the capture, or cross-sectional, area of the inlet. Additionally, there was a set of air-bleed doors on the fuselage, and a boundary layer fence was located in front of each intake next to the fuselage. Both the plugs and the doors were operated through hydraulic actuators. A set of auxiliary air inlet doors in the wheel wells were opened only when the landing gear was extended (this was the fix for the early belly landing problem). A lever and roller assembly, actuated by the main gear leg, closed the auxiliary inlets mechanically when the gear was retracted. During takeoff and subsonic acceleration, the plugs remained fully aft, allowing the maximum amount of air to the engine. At transonic speeds the VAI system was activated and controlled by the aircraft air data computer. Up to Mach 1.5, the bleed doors opened and closed depending on temperature and airspeed combinations, but above Mach 1.5 the bleed doors were kept open to allow potential shock waves to escape and not choke the inlet. The plugs moved forward according to a schedule determined by the Mach number and various data collected by the computer, and were fully forward by the time the aircraft reached Mach 1.92.

Tail Surfaces

All surfaces were highly-swept cantilever structures of aluminum and magnesium construction. Single-piece all-flying stabilators were connected by a hardened steel cross-beam, kinked downward to clear the engine tailpipe. The rudder used inset hinges with mass balancing at the top and hydraulic actuators at the bottom. No trim tabs were fitted, and a damper was included to prevent flutter at high speeds. A fixed magnesium ventral fin was fitted under the rear fuselage. All aircraft except the two YF-105As had an air intake at the base of the vertical stabilizer to provide cooling air to the aft fuselage and afterburner. This also provided a slight amount of additional thrust in the form of ejector airflow between the nozzle of the engine and the inner walls of the speed brakes.

Landing Gear

All models used a Bendix hydraulically retractable tricycle landing gear, with a single wheel on each unit. The main units retracted inward into thickened areas of wing-roots just aft of the main air-intake ducts, and the nose wheel retracted forward into the fuselage. Bendix oleo-pneumatic shock absorbers were fitted to both main and nose gears. During normal

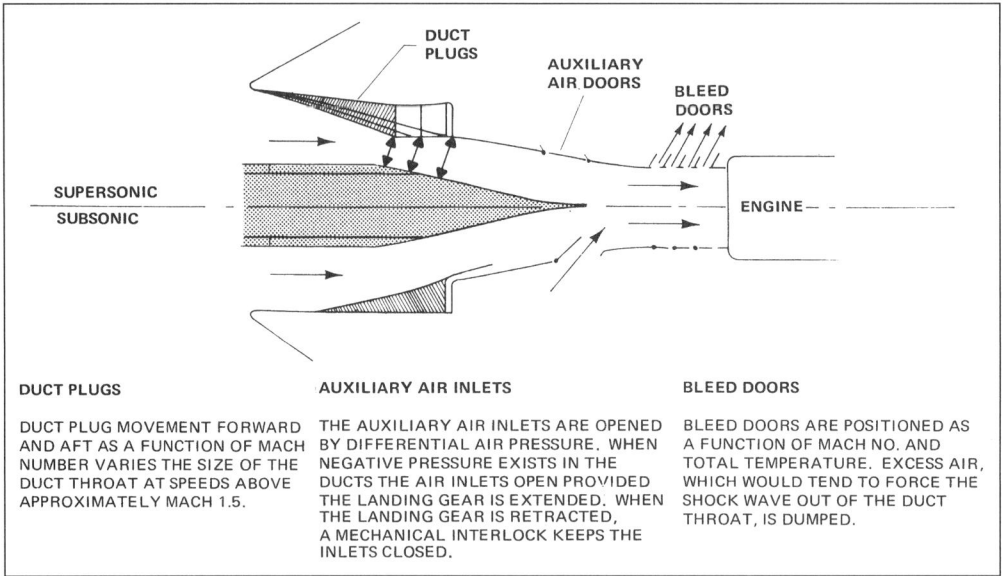

The variable air inlets of the F-105 adjusted the intake flow to engine needs through the use of intake duct plugs and auxiliary air doors. Excess air was removed through the bleed doors. Auxiliary air inlets inside the main landing gear wells opened when the gear was extended. The sequencing of these auxiliary inlets initially caused some problems at low airspeeds and high engine-rpm when the inlets opened and created sufficient suction to keep the landing gear from extending normally. *(U.S. Air Force)*

operation, the landing gear was retracted and extended by the utility hydraulic system. Retraction time was approximately 4–8 (3–7 in the F model) seconds, and extension required 5–9 seconds. The main gear was locked up by uplock hooks that engaged rollers on the inboard fairing doors. The nose gear was locked up by an uplock hook that engaged a roller on the nose gear strut. The uplock hooks were hydraulically actuated and contained an external spring to keep them in the locked position in case of a hydraulic failure. All gear were locked down by spring-loaded downlocks. The fairing doors remained open when the landing gear was extended. Emergency gear extension was possible by unlocking the uplock hooks with an emergency hydraulic accumulator and letting the gear drop to the down position by gravity. The main wheel tires were 36×11 inflated to 205 psi, and the nose wheel tire was 25×7.7 inflated to 140 psi. The main gear had multipad Goodyear hydraulic brakes equipped with antiskid units, except for early, unmodified B-models. An antispin system automatically applied the brakes during landing gear retraction to prevent fast spinning wheels from throwing debris into the wheel wells. A nose wheel steering button on the control stick grip engaged and disengaged the nose wheel steering system. This button served as the range scale button for the radar when airborne. Taxi lights were positioned below the radar reflector on the nose gear strut.

An arresting hook was installed on the aft fuselage centerline in the aft section of the ventral fin on F-105B-20-RE and subsequent aircraft, as well as early B-models that were processed through Look Alike. This system provided for runway overrun engagement with the BAK-6 water squeezer, BAK-9 brake system, or a modified MA-1A chain barrier. The self-centering hook assembly was attached to the aircraft structure by a pivot on its forward end and fitted with a spade-hook on its aft end. The hook was held into the up position by a shear pin. For hook extension, the pin was sheared by an electrically ignited explosive charge. When the pin was sheared, a cylinder combining the features of an accumulator and a shock strut extended the hook and held it in contact with the runway. The hook was not retractable in flight.

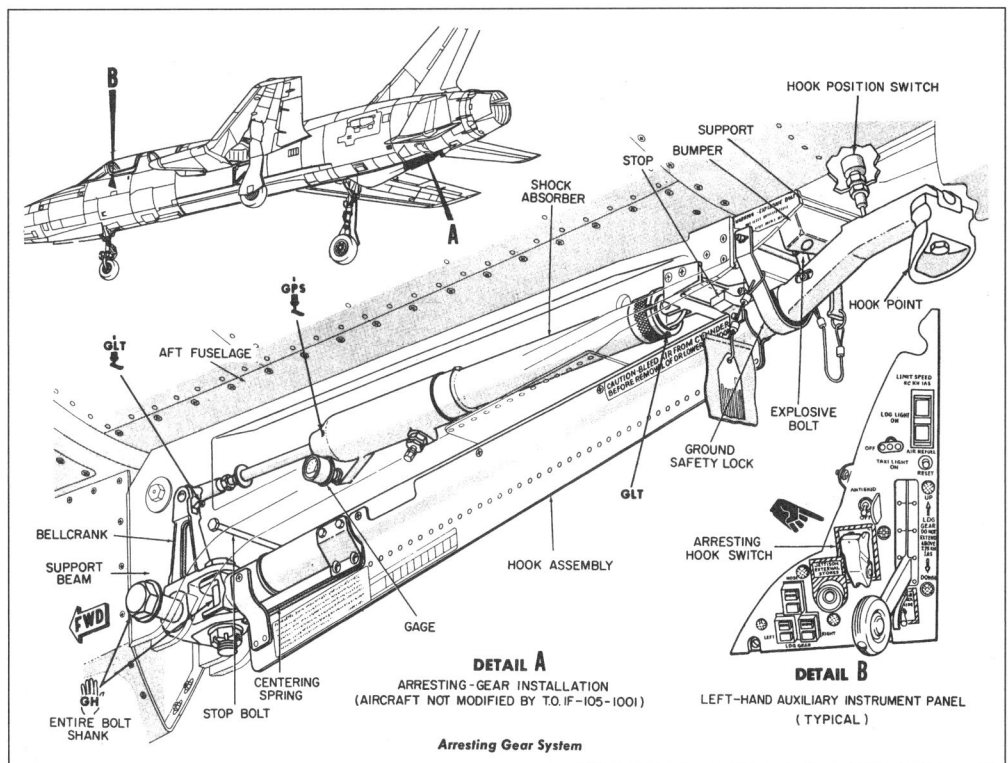

Beginning with block-20 B-models, an arresting hook was standard equipment. The hook was later retrofitted to block-10 and -15 B-models. The hook was never intended for use aboard aircraft carriers but was a useful addition for emergency landings at Air Force bases equipped with the appropriate cable systems. *(U.S. Air Force)*

Electrical System

All aircraft were equipped with an ac and dc power system. Primary system pushbutton-type circuit breakers were accessible from the cockpit. The 28 Vdc system was powered by a 400-ampere engine-driven generator with a 100-ampere battery as a standby source. Electrical power was distributed through a three-bus system consisting of a battery bus, a primary bus, and a secondary bus. The primary bus served equipment essential to flight and was energized by the generator, the battery, or the external power receptacle. The secondary bus served equipment not considered essential to flight and was energized by the generator or external source only. A fully charged battery was capable of providing the primary bus for 1.4 hours, extendible to 2.5 hours if only 10 minutes of command radio and IFF/SIF transponder time was used.

The ac power system was powered by a three-phase 115/200-volt, 400-cycle generator. Both the generator and the utility hydraulic pump were driven by an air turbine motor (ATM), which utilized air from the last stage of the engine compressor. In normal operation the ac generator distributed power through a two-bus system, identified as the primary and secondary buses. The primary bus used a step-down transformer to provide 26 volts to the instrument system.

Hydraulic System

The hydraulic system consisted of three independent systems—utility, primary-one, and primary-two. Each system had its own reservoir, pressure pump, pressure gauge, and plumbing. The utility system supplied hydraulic pressure to all systems of the aircraft except the primary flight control system. The primary-one and primary-two systems each operated one side of tandem actuators for the ailerons, spoilers, stabilators, rudder, and control stick boost. An emergency hydraulic system was provided to supply power to the primary flight controls in the event of engine failure. The emergency system consisted of a ram air turbine (RAT), hydraulic pump, and transfer valves, and utilized the utility system reservoir to provide hydraulic fluid under pressure to the primary-one system.

The utility hydraulic system provided hydraulic pressure to operate the landing gear, leading-edge flaps, speed brakes, refueling probe, bomb bay doors, nose wheel steering, wheel brakes, variable air inlet system, the gun drive, and water injection. Utility system pressure was maintained by a variable displacement pump driven by the ATM.

The control surfaces were operated by tandem actuators, each powered by both primary-one and primary-two hydraulic systems. If either of the primary systems failed, limited control of the aircraft was provided by the other system. If both systems failed due to engine failure, very limited control could be provided by extending the RAT and utilizing the emergency hydraulic system. The primary-one system also provided power for the automatic flight control system (AFCS). Each system was powered by an engine-driven hydraulic pump at approximately 3,000 psi. Airless, positive displacement type reservoirs were located in the main wheel wells. Air loads were not transmitted back through the controls, and artificial feel was provided by electric actuators, which were also used to trim the aircraft.

The normal hydraulic lines ran in the aircraft's belly, and relatively minor battle damage could cause complete loss of the hydraulic system, followed by the loss of the aircraft. Combat experience showed that the major cause in these cases was that without hydraulic pressure, the stabilators often slammed into a nose-high position, creating a sudden pitch-down that the pilot could not counter. In response, two modifications were incorporated that allowed the F-105 to be safely controlled and landed following a complete failure of the primary flight control hydraulic systems. The first was a pilot recovery mode (TCTO 1F-105-1045) that provided sufficient aircraft control without hydraulic pressure to allow a pilot to reach a safe area to eject. This mode consisted of a mechanical system that locked the stabilator in the neutral position when actuated by a switch in the cockpit. The pilot then controlled the aircraft in pitch by varying the thrust and by symmetrical deflection of the trailing-edge flaps, which were electrically controlled. Satisfactory control could be maintained up to Mach 0.85 and 20,000 feet, although the aircraft could not be very precisely controlled.

The second modification was the aircraft recovery system (TCTO 1F-105-1049), which used hydraulic pressure generated by the ram air turbine (RAT) in conjunction with a new hydraulic reservoir in the nose and a new set of hydraulic lines under a small fairing running the length of the upper fuselage. This system gave the pilot limited use of the stabilator and rudder and also allowed the landing gear to be extended for landing. No hydraulic pressure was provided to the ailerons or spoilers, although the unpressurized aileron actuators served as mechanical links and provided some control. The aircraft recovery system usually provided sufficient control for the pilot to make a safe return to a friendly base.

Miscellaneous Systems

A Hamilton Standard air-conditioning system operated when the engine was running. A duct directed hot, pressurized air to the air-conditioning package, the radar transmitter, the canopy seal, the anti-g suit, the external fuel tanks, and the dive shutoff valve through a heat exchanger. Temperature in the forward compartment was automatically maintained at approximately 85°F.

A 20-foot-diameter ringslot drag chute, generally used during landings, was packed in a deployment bag and stowed in an air-cooled compartment in the aft fuselage at the base of

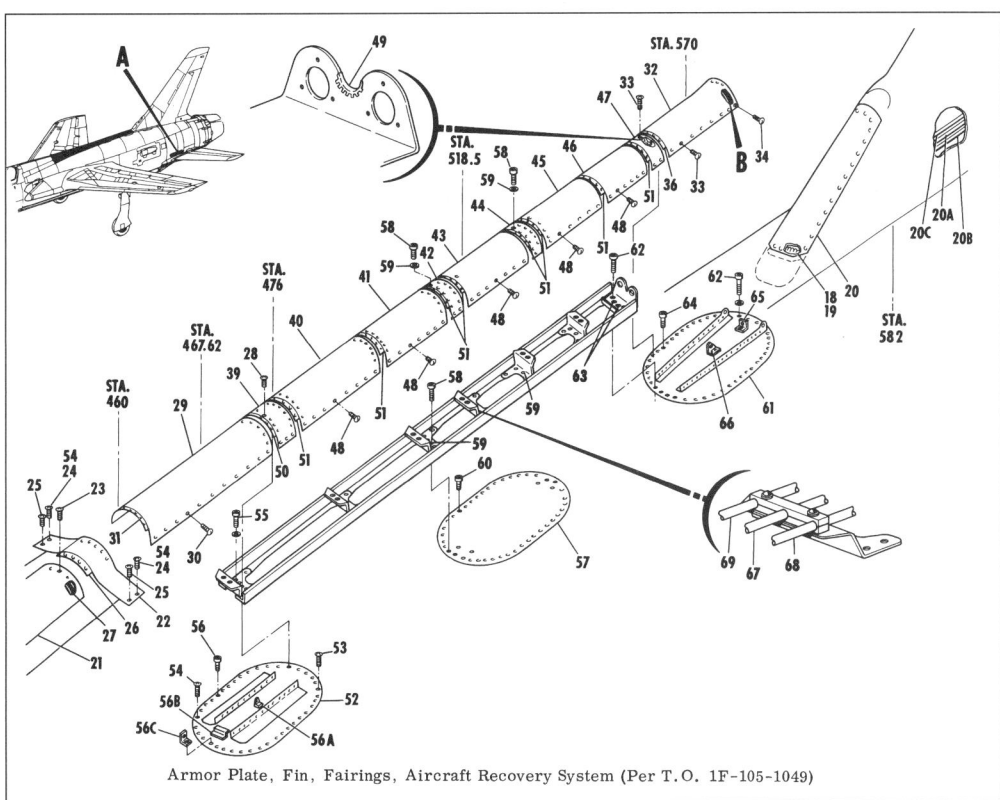

Armor Plate, Fin, Fairings, Aircraft Recovery System (Per T.O. 1F-105-1049)

The aircraft recovery system added a small fairing to the top of the fuselage that contained another set of hydraulic lines. These could be used to control the stabilators and rudder in the event the primary and secondary lines (which ran in the belly of the aircraft) were damaged. A small amount of armor plate was also added around the stabilator actuators. *(U.S. Air Force)*

the rudder. A nylon riser connected the drag chute to the jettison hook on the aircraft. A shear pin was provided as a safeguard against the drag chute being deployed at too high a speed, the pin designed to fail under loads above 200 knots, thereby releasing the chute without damaging the aircraft or making it uncontrollable.

Avionics

The AN/ASQ-37 communications, identification and navigation (CIN) set for the F-105 series included the AN/ARC-70 UHF command radio, AN/ARA-48 ADF group, AN/ARN-61 instrument landing system, AN/ARN-62 TACAN, AN/ARN-61 marker beacon, and AN/APX-37 IFF/SIF transponder. Later additions would include the AN/ARC-164 UHF/VHF command radio, while the APX-37 would be replaced with the AN/APX-72 IFF set.

The original F-105B fire-control system was the General Electric MA-8. It was an integrated assembly of GE equipment, including an E-34 ranging radar, E-50 gun-leading computing sight system, E-30 toss-bomb computer, and a T-145 weapons system for nuclear weapons. A Bendix central air data computer provided altitude, air density, and temperature information to a General Electric FC-5 autopilot that enabled the F-105B to deliver nuclear weapons during daylight conditions but was not particularly effective with conventional weapons. An APN-105 Doppler navigation set was also fitted. The AF/A42G-1 (XK-5) automatic flight control system (AFCS) provided F-105B aircraft with two basic modes of operation: stability augmentation and pilot relief. The stability augmentation mode improved control of the aircraft by damping oscillations about its pitch and yaw axes. Automatic turn coordination was provided to counteract sideslip or skid. The pilot

A good visual display of the components of the MA-8 fire control system used in the F-105B. At the time of its development the MA-8 was the most complex system ever fitted to a fighter aircraft, but by the time it was deployed it was already considered obsolete, leading to the development of the ASG-19 Thunderstick system for the F-105D/F. *(San Diego Aerospace Museum Collection)*

relief mode provided roll stability augmentation, altitude and Mach hold, attitude and heading hold, and track-hold functions.

The major advancement of the D-model (and also used in the F/G) was the ASG-19 Thunderstick integrated fire control system. This provided the pilot the option of visual or blind weapons delivery for nuclear or conventional weapons, as well as an air-to-air mode. The system could also provide ground mapping, contour mapping, and terrain avoidance functions. The ASG-19 included the R-14A NASSAR (North American Search and Ranging Radar) that had been developed by the Autonetics Division of North American Aviation. This radar had search and ranging functions available in either the air-to-ground or air-to-air modes. It was of monopulse type and operated in X-band for improved detection and jam resistance. The radar scope consisted of a 5-inch cathode ray direct view storage type tube on the lower center of the instrument panel. The APN-131 Doppler navigation unit was a self-contained

Looking a little beat-up, this was the antenna for the R-14A radar used in the F-105D/F. This aircraft was being parted-out at MASDC, explaining its condition. *(Mick Roth)*

microwave transmitter-receiver that could present great-circle courses to any point on the globe, and presented visual displays of present position, ground speed, and drift angle. The APN-131 was built by Laboratory for Electronics, Inc., of Boston, Massachusetts.

The AF/A42G-8 (FC-5) AFCS provided F-105D/F aircraft with three basic modes of operation: stability augmentation, pilot relief, and fully automatic. The stability augmentation and pilot relief modes were similar to those provided by the F-105B's AFCS. The fully automatic mode enabled the aircraft to make automatic instrument approaches to a runway, automatic bomb-toss maneuvers, and to maintain wings-level on a bombing run. In this mode, the control stick moved in conjunction with the flight control surfaces.

The Thunderstick-II system was intended to give the F-105D a true blind-bombing capability, as well as to improve its visual bombing from low altitudes. A new ITT ARN-92 LORAN receiver was added in the turtledeck fairing that added 22 cubic feet of volume, made no significant change in drag, and actually improved directional stability somewhat. The LORAN antenna was taped to the inside of the canopy, just over the pilot's head. The R-14A radar system had its original electronics replaced with solid-state devices and was redesignated R-14K. A Singer/General Precision gyro-compassing attitude vertical reference system worked in conjunction with the APN-131 Doppler navigation unit.

Paint and Markings

All early F-105s came from the factory in a natural metal finish with flat black antiglare paint extending along the cockpit fairing almost to the base of the vertical stabilizer. A few aircraft destined for test programs were painted an overall light gray. Project Look Alike

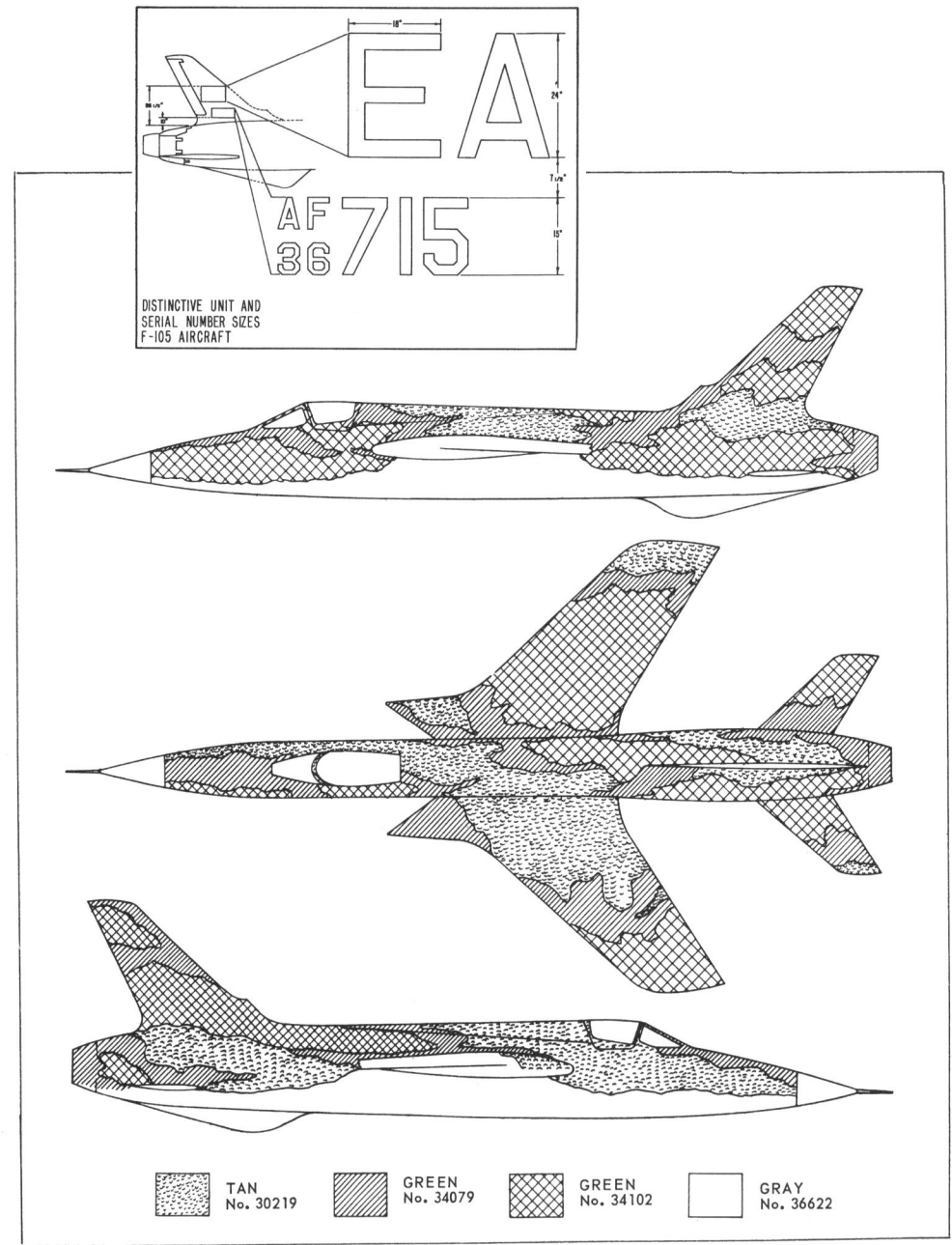

F-105 Aircraft Camouflage Pattern

Beginning in late 1966, F-105s started wearing the standard Vietnam-era camouflage of tan (FS-30219) and two shades of green (FS-34079 and FS-34102) on top, with light gray (FS-36622) on the bottom. Approximately one aircraft in every four had the tan and 30219 green reversed. Approximately 20 gallons of acrylic nitrocellulose lacquer were used on each aircraft. *(U.S. Air Force)*

introduced an aluminized silver acrylic-lacquer in an attempt to better seal the aircraft against corrosion. Again, there was flat black antiglare paint. All F-105s had been painted by June 1964.

Beginning in 1966, F-105s started wearing the standard Vietnam-era camouflage of tan (FS-30219) and two shades of green (FS-34079 and FS-34102) on top, with light gray (FS-36622) on bottom. Approximately one aircraft in every four had the tan and 30219 green reversed. The aircraft were repainted as they passed through the normal IRAN

A J75 engine installed in an F-105B. Even though the B-models did not serve in Southeast Asia, they received the large cooling air intake scoops along the sides of the fuselage. The J75 engine was exceptionally large, and for its day, very powerful. Note how the ventral fin splits when the rear fuselage is removed. The arresting hook is on the portion of the ventral fin that is permanently attached to the rear fuselage. *(Mick Roth)*

process. Since this was spread over a considerable time, many F-105s were initially seen in Southeast Asia still finished in their Look Alike silver. Approximately 20 gallons of acrylic nitrocellulose lacquer were used on each aircraft.

After their return from Southeast Asia, F-105s were painted in several camouflage schemes. One of these was known as Desert Fox, and was intended for use in desert environments. Another was a variation of the European One scheme of green and dark gray. Most F-105s, however, finished off their careers in a wraparound version of the standard Southeast Asia camouflage that simply continued the green/tan onto the undersides of the aircraft, replacing the earlier light gray.

Armament

The F-105's fixed armament consisted of one General Electric M61 (earlier called a T-171E-3) 20-mm Vulcan rotary cannon. The cannon had six barrels, weighed 275 pounds, was hydraulically driven, and fired through a port on the lower left side of the fuselage. Muzzle velocity was 3,380 feet per second. Although the original specification called for 1,130 rounds of M-50 series electrically primed 20-mm ammunition, aircraft generally were not loaded with more than 1,028 rounds. This furnished approximately 10 seconds of firepower, although the normal burst was only 3 seconds. Republic developed two different feed systems for the cannon. In the F-105B it was a dual feed system, with two feed chutes guiding the shells into the breech. The links and empty shell casings were retained after firing and were stored in separate compartments. The rationale for this arrangement was to minimize the effects of firing 500 pounds of ammunition on the center of gravity. The F-105D required additional space in the nose for the new radar system, and the only volume available was the link and shell storage area. The feed system was redesigned with a linkless belt moving the shells from their storage drum to a conveyer that fed them to the breech. Empty shell casings were extracted from the breech and fed back into the drum for storage. This system recovered about 16 cubic feet.

The aircraft was fitted with a 16-mm camera mounted on the gunsight. The camera had a 50-foot magazine and the recording rate was adjustable between 16, 32, and 64 frames per second. All F-105Bs and aircraft through block-10 D-models were fitted with a KB-3A cam-

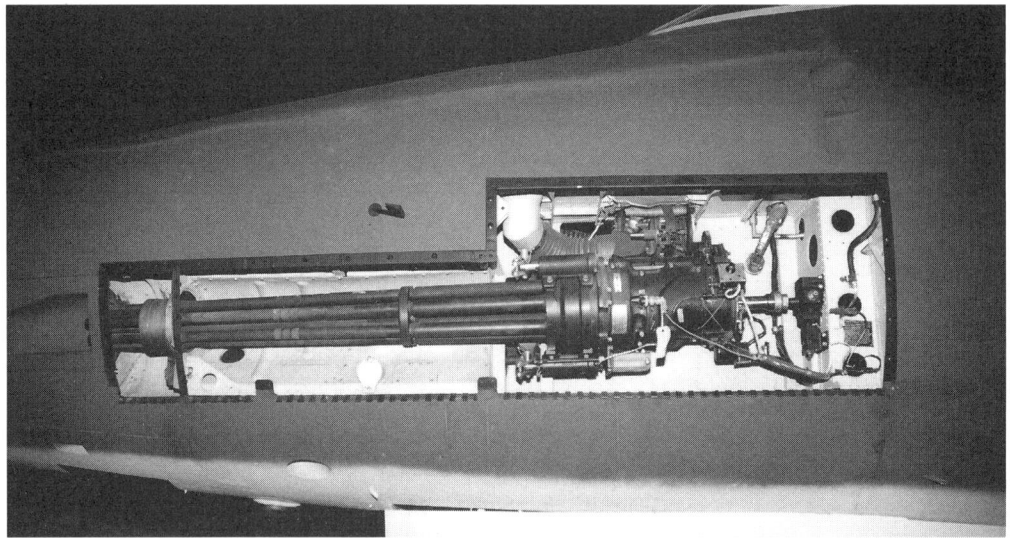

All F-105s were armed with a single General Electric M61 Vulcan 20-mm cannon in the nose. This is the installation in an F-105D-5-RE (58-1155), as currently shown in the Air Force Armament Museum at Eglin AFB, Florida. This particular F-105D is exceptionally well preserved. *(Dennis R. Jenkins)*

era, which required manual exposure control. Starting on block-15 D-models, a KS-27A or KS-27B equipped with automatic exposure control was used.

The aircraft had an internal weapons bay in the fuselage, under the wings, capable of carrying one nuclear weapon. When the weapons bay doors were opened, there was no trim change to the aircraft. Internal stores could be dropped without encountering any buffeting or stability loss. The bomb bay was also capable of carrying a 390-gallon fuel tank. The fuel tank was a fixed installation and could not be jettisoned.

All versions were also capable of carrying a 450- or 650-gallon centerline tank, 450-gallon tanks on inboard wing pylons, and Maxson (Martin) GAM-83 (AGM-12) Bullpup missiles or GAR-8 (AIM-9) Sidewinder missiles on the outer wing pylons. Some of the fuel tanks were constructed integral with their pylons; these can be identified by the absence of sway braces on the pylon. In all cases, the pylons, and any stores on them, could be jettisoned from the aircraft.

Powerplant

The powerplant used in the two YF-105As was the Pratt & Whitney J57-P-25. It was the same basic engine as the J57-P-21 used in the F-100, with two detail changes. First, the engine exhaust nozzle was extended by 22 inches into an afterburner flap-type nozzle, and second, the afterburner control unit was mounted on the side of the engine. Only six of the engines were ever assembled, and they were all used to support the YF-105As. The engine developed 16,000 lbf under sea-level static conditions. Its military rating was 10,200 lbf, and 90 percent cruise thrust worked out to 7,800 lbf. The J57 series featured two compressors stages in series, with a drive turbine attached to each. The nine-stage low-pressure compressor was driven by a two-stage turbine, while the seven-stage high-pressure compressor was driven by a single-stage turbine. The engine was 155 inches long without the afterburner, 39 inches in diameter, and weighed approximately 4,200 pounds.

Production aircraft used one Pratt & Whitney J75 (JT4A) turbojet engine of various dash numbers, depending on the F-105 model. The J75 is an axial-flow turbojet with eight low-pressure (LP) and seven high-pressure (HP) stages, a two-spool compressor, an annular combustion chamber with eight flame tubes, each with six fuel nozzles, a three-

Ground crew install a loaded ammunition drum into the nose of a 333rd TFS F-105D at Takhli in 1965. The ammunition drum held a total of 1,028 rounds of 20-mm ammunition—less than a 10-second supply at 6,000 rounds per minute (the standard burst was 3 seconds). The ram air turbine is extended just under the ammo drum. *(U.S. Air Force)*

stage (one HP and two LP) turbine, and an afterburner with a two-position exhaust nozzle. The compressor rotor assemblies are mechanically independent of each other. The high-pressure compressor rotor is connected to, and driven by, the first-stage turbine wheel via a hollow shaft. A shaft rotating within the hollow shaft independently joins the low-pressure compressor rotor to the combined second- and third-stage turbine wheels. Control, fuel, and lubrication system were generally similar to the J57 series engines. The J75-P-19W as used in the F-105D/F/G (as well as the F-106) had a pressure ratio of 12:1, and was 20 feet long, 43 inches in diameter, and weighed approximately 5,000 pounds. Oil capacity was 6.5 gallons.

With maximum afterburning and water injection, the J75-P-19W developed 26,500 lbf[214] at sea level on a standard day. During each minute it consumed 15,720 pounds of air and 972 pounds of JP-4. Its military rating was 16,100 lbf, with a fuel burn reduced to 220 pounds per minute. A sensor in the engine air ducts detected icing conditions and automatically ducted hot bleed air into the ducts for anti-icing. On F-105Bs and block-5 and -10 D-models, the probe was not positioned correctly, and flight into known icing conditions was prohibited until this was corrected by TCTO 1F-105-630 and 632.

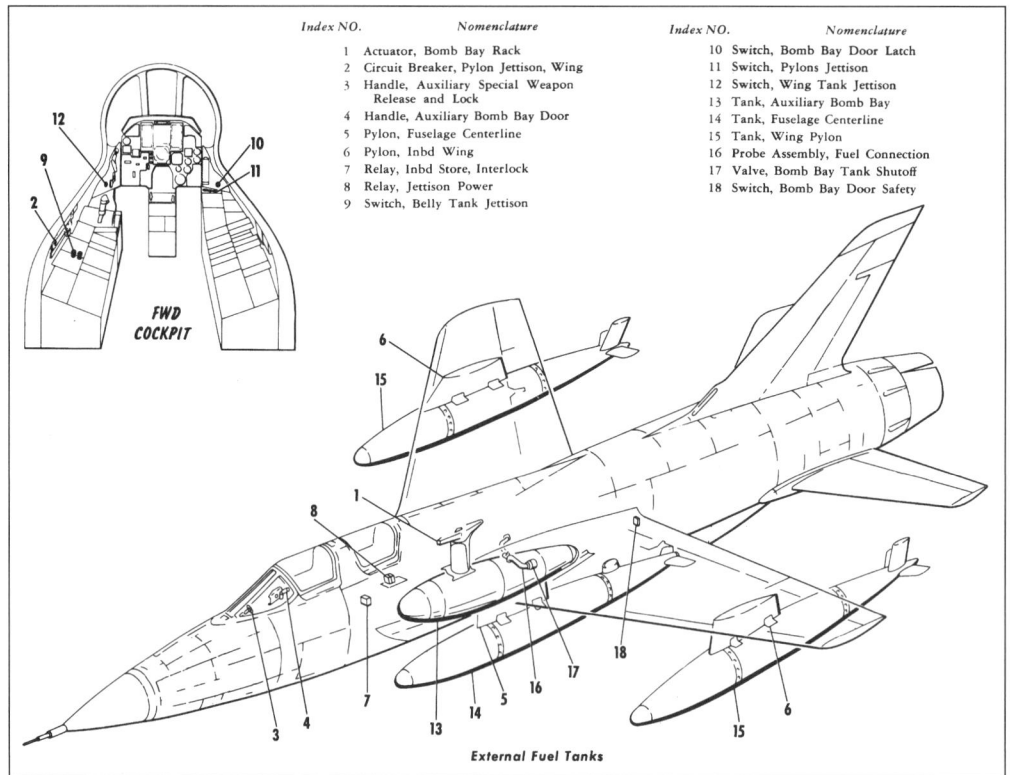

Up to four auxiliary fuel tanks could be carried by the F-105. In Southeast Asia almost all F-105s carried the 390-gallon weapons bay fuel tank. The use of the 450-gallon wing tanks and 450- or 650-gallon centerline fuselage tank depended upon the weapons configuration being carried on a particular mission. Some of the fuel tanks were constructed integral with their pylons and can be identified by the absence of sway braces on the pylon. *(U.S. Air Force)*

This F-105D is carrying an AIM-9 Sidewinder on the outer wing pylon and a MER with three M117 750-pound bombs on the inner pylon. Due to interference with the main landing gear, the MER on the inboard pylon could only carry four weapons instead of six (the two inboard stations were empty). *(Larsen/Remington via the Mick Roth Collection)*

F-105s often used AGM-12B Bullpups in Southeast Asia, but they were difficult to use since the pilot had to provide guidance, leaving him exposed to ground fire. Interestingly, AGM-12C Bullpups were still being used in 1970. The missiles were primarily used against caves, although some were also fired at roads, bridges, and trucks. Bullpups were normally released in a 20° dive at about 18,000 feet slant-range, or somewhat closer. *(Cradle of Aviation Museum via Ken Neubeck)*

The Maxson (Martin) GAM-83 (AGM-12) Bullpup missile could be carried on the outer wing station. Provisions for launching the conventional version (mainly wiring) were included beginning on the 70th F-105D, but the actual equipment to do so did not appear until the 115th D-model. This aircraft also included the capability to carry the nuclear-armed version of Bullpup, and these capabilities were later retrofitted to all F-105Ds. *(San Diego Aerospace Museum Collection)*

The last JF-105B-2-RE (54-0112) was used for a variety of weapons carriage/separation tests. Note the unusual checkerboard pattern on the wing. An ALE-2 chaff pod is shown on the outer wing pylon. Note the ram air duct in the front of the chaff pod and the slot in the top of the pod just behind the pylon where the chaff came out. *(Mike Machat Collection via Tony Landis)*

On the F-105B, fuel was carried in seven separate fuel tanks lined up along the upper rear fuselage behind the pilot. Total internal fuel capacity was 900 gallons, of which 25 gallons of useful fuel was actually located in the fuel lines. Two 450-gallon drop tanks could be carried under the inboard underwing pylons, and an additional 450- or 650-gallon drop tank could be carried on a hardpoint underneath the fuselage. In addition, the weapons bay

A dual AIM-9 Sidewinder launcher was also tested on the last JF-105B. *(San Diego Aerospace Museum Collection)*

The Details—Construction and Systems

CONVENTIONAL MUNITIONS F-105B/D/F AIRCRAFT

T.O. 1F-105B-33-1-1
circa 1971

STORE	RIGHT OUTBD			RIGHT INBD					CENTER							LEFT INBD					LEFT OUTBD		
	UNIVERSAL PYLON	AGM ADAPTER	AIM ADAPTER	PYLON/MER	MULTIPLE WEAPON PYLON	UNIVERSAL PYLON	AGM ADAPTER	FUEL TANK PYLON	MULTIPLE WEAPON PYLON WITH MER	MULTIPLE WEAPON PYLON	MER PYLON	MER PYLON WITH MER	F-105B FUEL PYLON	FUEL TANK PYLON	M-61 GUN	FUEL TANK PYLON	AGM ADAPTER	UNIVERSAL PYLON	MULTIPLE WEAPON PYLON	PYLON/MER	AIM ADAPTER	AGM ADAPTER	UNIVERSAL PYLON
M-61 (Linkless Feed) (F-105D/F)															1								
M-61 (Dual Feed) (F-105B)															1								
M-117 (750 LB GP Bomb)	1			4	1	1					6							1	1	4			1
M-117 (Retarded)	1			4	1	1					6							1	1	4			1
M-117D (Destructor)	1			4	1	1					6							1	1	4			1
MK-36 (Destructor)	1			4	1	1					6							1	1	4			1
MK-82 (500 LB GP Bomb)	1			4	1						6								1	4			1
MK-82 (Snakeye)	1			4	1	1					6							1	1	4			1
MK-83 (1000 LB GP Bomb)	1			2	1	1					3							1	1	2			1
MK-84 (2000 LB GP Bomb)				1	1													1	1				
M-118 (3000 LB GP Bomb)				1	1													1	1				
M-129-E1, E2 (Leaflet Bomb)	1			4	1	1					6							1	1	4			1
MC-1 (750 LB Gas Bomb)	1			4	1	1					6							1	1	4			1
BLU-1/B (Fire Bomb)	1			2	1	1					2							1	1	2			1
BLU-1B/B (Fire Bomb)	1			2	1	1					2							1	1	2			1
BLU-1C/B (Fire Bomb)	1			2	1	1					2							1	1	2			1
BLU-27/B (Fire Bomb)	1			2	1	1					2							1	1	2			1
BLU-52/B, BLU-52A/B (350 LB Chem. Bomb)	1			2	1	1					3							1	1	2			1
AIM-9B, AIM-9E (Missile)			2																		2		
TDU-11/B (Rocket)			1																		1		
AGM-12B (Missile)		1										1										1	
AGM-12C (Missile)							1										1						
AGM-45/A (Missile)		1																				1	
AGM-78 (Missile)							1										1						
LAU-3/A (Rocket Launcher)	1				2														2				1
LAU-32A/A (Rocket Launcher)	1				2														2				1
LAU-32B/A (Rocket Launcher)	1				2														2				1
LAU-59/A (Rocket Launcher)	1				2														2				1
CBU-1A/A	1																						1
CBU-2/A	1																						1
CBU-2A/A	1																						1
CBU-2B/A	1																						1
CBU-2C/A	1																						1
CBU-3/A	1																						1
CBU-3A/A	1																						1
CBU-7/A	1											1											1
CBU-9/A	1																						1
CBU-9A/A	1																						1
CBU-9B/A	1																						1
CBU-12/A	1																						1

This is the conventional munitions compatibility chart for the F-105 in 1971. This chart does not show ECM pods, fuel tanks, or nuclear weapons. Note that the B-model is included in the chart and can carry all of the weapons listed for the later D/F models. *(U.S. Air Force)*

could accommodate a 390-gallon fuel tank in place of weapons. Total fuel load (internal plus external) could be as high as 2,865 gallons.

On the F-105D and F, fuel was carried in three bladder-type tanks in the fuselage (forward, main, and aft) with a total capacity of 1,135 gallons. Provisions existed for one 390-gallon bomb bay tank, either a 450- or a 650-gallon centerline tank, and a 450-gallon tank

The 20-mm cannon muzzle on the F-105B overlapped the rear part of the radome. The entire forward fuselage would be redesigned for the D/F-models, dictated largely by the need to incorporate a much larger radar unit for the ASG-19 fire control system. *(Mick Roth)*

The business end of a J75 engine installed in an F-105D. Note the four-petal speed brake surrounding the exhaust. *(Dennis R. Jenkins)*

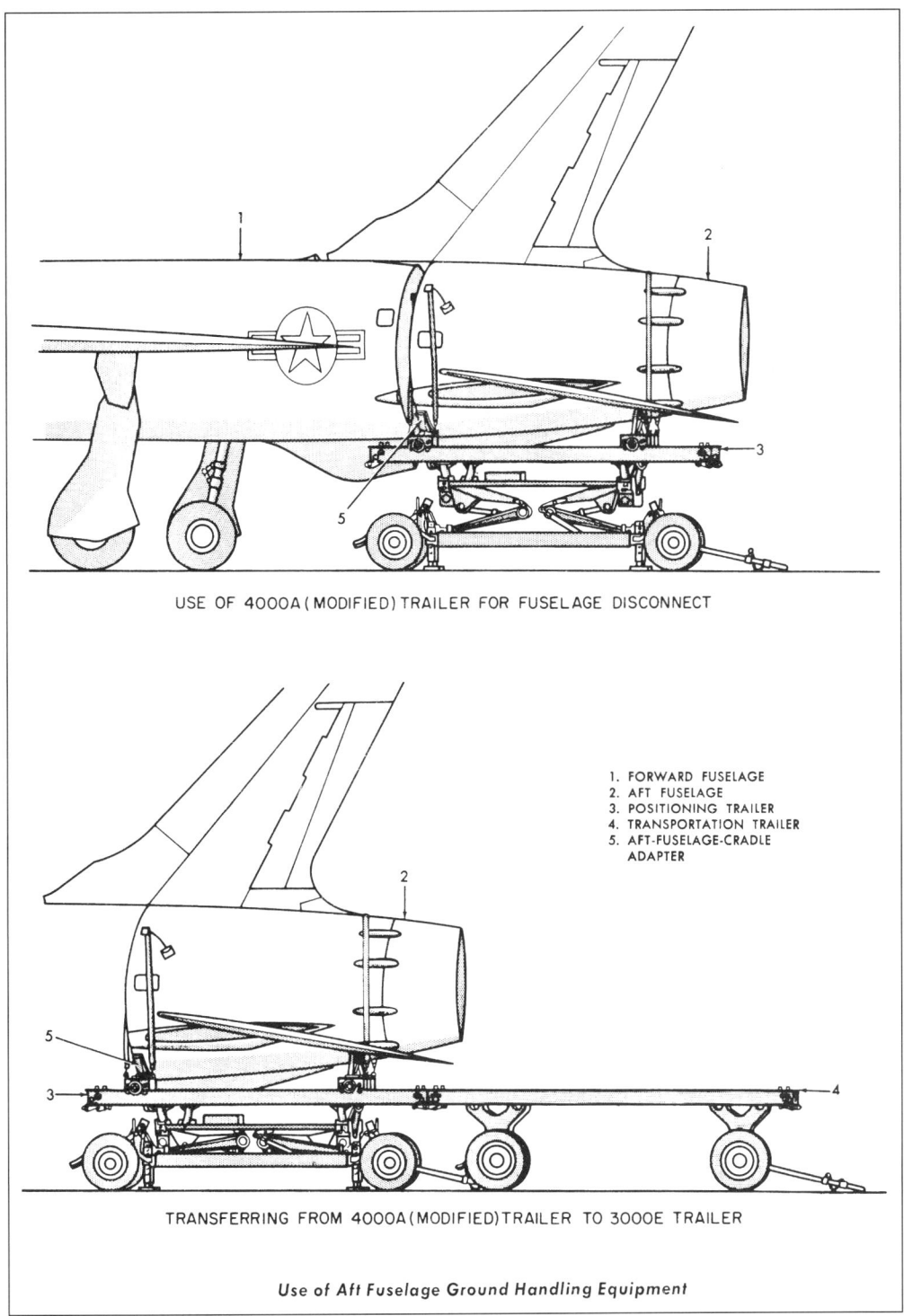

Use of Aft Fuselage Ground Handling Equipment

A hydraulic trailer was used to support the rear fuselage while it was detached from the rest of the aircraft. The rear fuselage was then transferred to another trailer. The hydraulic trailer was then used to support the engine while it was removed from the aircraft. *(U.S. Air Force)*

to be carried on each inboard wing station. The weapons bay tank was not jettisonable, and some models of the centerline and wing tanks were integral with their pylons. The arrangement of internal tanks differed somewhat among the models and block numbers. As an example, early F-105Ds (block-5 through block-25) carried 365 gallons in the forward tank, 251 gallons in the main tank, and 519 gallons in the aft tank. Later block-30 and block-31

All F-105s were originally equipped with a probe-and-drogue in-flight refueling system using this fully retractable probe. The last batch of F-105Ds and all F-105Fs were also equipped with a boom-type refueling receptacle on top of the fuselage ahead of the windscreen. All earlier F-105Ds were retrofitted with the boom receptacle as part of their Project Look Alike modifications *(Cradle of Aviation Museum via Ken Neubeck)*

aircraft carried 376, 257, and 502 gallons respectively. Total fuel load (internal plus external) could be as high as 3,100 gallons.

Like many aircraft of its era, the F-105 had one design feature that proved to be very susceptible to damage. The engine was surrounded on three sides by a saddle fuel tank. If the engine came apart, either through a failure of the engine itself or as a result of battle damage, chances were good that pieces of the engine would penetrate the saddle fuel tank, resulting in a catastrophic explosion. There was nothing that could realistically be done about this, but designers learned a critical lesson, and most current fighters do not carry fuel around the engine compartment.

The engine drew fuel from the main fuselage tank, all other tanks feeding it in sequence. The tanks had been designed with a built-in explosion-detection and suppression system, but the Air Force deleted it prior to production to save weight and costs. After several bad experiences in combat over Southeast Asia, self-sealing fuel cells with foam were retrofitted (TCTO 1F-105D-1058/59) to surviving aircraft in an extremely costly modification program. A single-point refueling receptacle was located on the left fuselage, aft of wing.

All F-105s had a retractable probe-and-drogue refueling probe located on the upper left side of the fuselage just forward of the canopy. F-105D-31-REs and all F/G models were also equipped with a flying-boom receptacle slightly forward of the probe, making the Thunderchief one of the few U.S. combat aircraft to be equipped with both systems simultaneously. Most earlier F-105Ds were eventually retrofitted with the receptacle system.

Notes

210. *Introduction to the F-105F Two-Place Airplane,* Republic Aviation Corporation, undated (probably early 1963), p 4–9.
211. *Ibid.,* p 10.
212. *Ibid.,* p 3.
213. *Ibid.,* p 3.
214. Without water injection the engine developed 24,500 lbf in afterburner.

APPENDIX A

Fire Control System Development

The F-105 was developed during a period of tremendous technological advances, and the MA-8 fire control system attempted to integrate a variety of complex electromechanical and electronic systems to provide the F-105 with a true all-weather capability. It should be noted that the Air Force was not a monolithic entity. Airframe development was initiated by various aircraft program offices in concert with the eventual user Command, while armament development was handled by the Armament Laboratory. These organizations often treated each other as though they were direct competitors, or even enemies. Communication between them was generally in the form of requirements documents and official memorandums instead of cooperative agreements, which sometimes led to a very confusing environment. It should also be noted that requirements issued by Wright Field for theoretical aircraft did not necessarily bear any resemblance to actual aircraft they may, or may not, have intended to build.

The MA-8 program began in April 1951 when the Air Force Armament Laboratory began looking at using a three-gyroscope stabilized computing sight to form the basis of an advanced fighter-bomber fire control system. The Sperry Gyroscope Company received a contract on 7 December 1951 to develop an advanced sight that could be integrated with an air-to-ground radar. On 5 January 1952 the Air Force published a general operational requirement (GOR 52-1) for the fire control system destined for an unnamed 1960 fighter-bomber whose primary mission would be the destruction of fixed or moving ground targets in any weather. Since the first control system also had to allow the aircraft to defend itself against enemy fighters, it needed to have an air-to-air capability. The bombing mode of the system was to allow both high- and low-altitude visual attacks using nuclear weapons with a "high probability of target destruction."[1]

On 22 January 1952 a set of military requirements had been issued aimed at a 1956 fighter-bomber capable of performing visual dive-bombing and low-level air-to-ground attacks during daylight, operating from 5,000-foot landing strips in forward areas, and defending itself against air attack ". . . as a last resort." Potential armament included 2.75-inch and 5-inch unguided rockets and conventional bombs.[2]

The MA-4 configuration included the new Sperry A-5 optical sight, Sperry AN/APG-31 air-to-ground ranging radar, M-1 toss bomb computer, MA-1 bombing computer, a time-of-flight computer, and a power supply. However, by July 1952 General Electric had proposed a new version of its K-19[3] optical sight that provided toss bombing computations as well as partial computation for air-to-ground rockets. These capabilities would allow an "interim" configuration, designated MA-8, to be developed and fielded much faster than waiting for the new Sperry sight to be completed. Both automatic and manual bomb release were required at dive angles ranging from 0 to 85°. Computations for lead pursuit air-to-air gunnery and rocket fire, and for air-to-ground bombing and rocketry, were required. The sight gyro would be caged for air-to-ground gunnery. When the Sperry A-5 sight had completed its development it would replace the interim K-19 unit. There would be another major difference between the two systems. In the interim system all of the components would be connected and "lose their identity," but there would be a considerable amount of functional overlap. In the final MA-4 system, all the components would be fully integrated and most of the overlap would be eliminated.

On 4 May 1953 the Air Force proposed that Republic take over the responsibility for the interim MA-8 fire control system and subcontract the final development to a manufacturer "not objectionable to the Armament Laboratory." A conference on 25 June tentatively approved the use of the Sperry A-4 sight (instead of the GE K-19), MA-4 time-of-flight computer, and M-1 toss bomb computer. The MA-1 bombing computer would still be installed as a separate item and would not form an integral part of the MA-8 system.

On 30 July 1953 it was decided to accept a General Electric proposal to complete development of and manufacture the MA-8, but using the improved GE K-19 sight instead of the Sperry unit. One factor in GE's favor was a proposed integrated computer that could perform the functions of both the M-1 and MA-1. This would eliminate the need for a separate bombing system built around the MA-1. Furthermore, the new computer was to be capable of directing the aircraft on an almost unlimited number of delivery paths. When the pilot located his target visually and locked on, he would fly straight until receiving an indication from the system to pull up. During the pull-up the bomb would be released automatically if a hit was possible. This would be a tremendous improvement over the narrow bomb drop pattern of the MA-2. However, the new computer existed only on paper, and the M-1 and MA-1 were kept in mind as possible contingency replacements.

On 15 October 1953 the Air Force officially canceled the MA-4 development program and also terminated the Sperry A-5 sight. This was largely because the interim MA-8 with the new GE computer was expected to have sufficient growth potential to cover any expected uses, and the Massachusetts Institute of Technology was developing an even more advanced sight under the classified Black Warrior program.

By the middle of 1954 the new GE computer had run into trouble. General Electric had been working on the theory of the device for several years prior to proposing it for the MA-8, but the task of turning the design into actual equipment was proving more difficult than expected. Instead of delivering the first MA-8 development system in May 1954, it now appeared more likely to happen in November, at the earliest. Furthermore, the computer continued to grow in size, almost doubling from the original 1,280 cubic inches to 2,494 cubic inches.

The Tactical Air Command also complicated matters by issuing new requirements. In March 1955 TAC requested that the F-105 be equipped for all-weather operation, and in addition, that it include an autopilot tied in to the fire control system. TAC wanted the autopilot to provide not only automatic bomb delivery, but also automatic safe recovery from bomb-drop maneuvers. On 26 September an "airborne data line receiving system" was approved for all tactical aircraft so that they would be compatible with the "1957–60 Tactical Air Control System Environment," something conceived to be generally similar to the Semi-Automatic Ground Environment (SAGE) being developed for the Air Defense Command.

Each new requirement meant adding more equipment into an already crowded airframe. This in turn forced the autopilot, fire control system, and electronic component manufacturers to reduce the size and weight of their products, further delaying their production. At this point the XMA-8 contained a modified K-19 optical sight, a Kollsman air data computer (to measure angle of attack, true airspeed, and air density), a modified Sperry APG-31 radar, AN/APW-11A beacon transmitter, and either the new GE computer or the M-1/MA-1 combination. Another computer was also being considered as a contingency. The AC Spark Plug AC-13 was being developed for the XMA-12 in the F-107 and could be used if its development continued on track.

Flight testing of the XMA-8 finally got underway aboard a modified RF-84F (51-1835) during October 1955. On 15 February 1956 the Air Force met internally to discuss the requirement to tie the autopilot into the fire control system. The Flight Control Laboratory held that automatic control of the roll axis to maintain a wings-level attitude for target acquisition to the bomb release point was feasible, but that pitch control was not necessary. However, the control system could be constructed so that it would guide the aircraft through a single type of preprogrammed escape maneuver.

A funding reduction in FY57 prompted the Air Force to concentrate the RF-84F testing on the bombing mode of the MA-8. It was already obvious that the F-105 was destined to be a true fighter-bomber, and any air-to-air work or strafing missions would be purely secondary. Reductions in gun and rocket firings also seemed appropriate since the K-19 had been extensively tested during other projects.

By May 1957, General Electric had finally demonstrated its new bombing computer, eliminating the M1/MA-1 and AC-13 from further consideration. Shortly thereafter the GE plant in Schenectady, New York, began turning out small numbers of MA-8s for use in the F-105B. In its production configuration, the system used or interfaced with an E-34 radar ranging set, E-50 gyroscopic sight, E-30 toss-bombing set, a nuclear weapon safing and arming system, Bendix navigation computer, and General Electric FC-25 autopilot. The MA-8 could accommodate the use of the TX-28[4] nuclear weapon and 750-pound bombs carried in the internal bomb bay, four AIM-9 Sidewinders, ninety-five 2.75-inch FFAR rockets, and the internal T-171E-3 20-mm cannon. Work was underway, for some odd reason, to integrate the MB-1 Genie into the system, although it is unclear as to whether this ever actually happened.

By the end of 1958 fire control system tests had revealed a number of problems, mostly concerning the sight reticle. The reticle was raised so that the pilot did not have to lean forward to use it, its brightness increased, jitter adjusted, and an automatic reticle uncaging unit added to simplify air-to-air gunnery. In addition, "vertical gyro fast erection" circuitry was added to cut gyro errors and changes were made to eliminate excessive radar waveguide breakage.

Operational testing of the MA-8 in the F-105B was unveiling both reliability and accuracy problems. In addition, the system was exhibiting more all-weather limitations than expected. A number of modifications were developed, and the system eventually met its original requirements, but not before TAC had decided it needed a more capable system.

GOR-49-1, which essentially defined the requirements for the follow-on F-105D, called for an even more advanced fire control system. General Electric was again selected to develop the system incorporating the R-14A NASARR, a new optical sight, a transistorized version of the bombing computer, and an offset initial point bombing computer. To use the offset bombing device, the pilot had to have an initial point and know its distance and azimuth with respect to an initial point, but did not have to overfly the initial point directly. Republic called the system Thunderstick, and at the end of 1958 the new fire control system was officially designated AN/ASG-19.

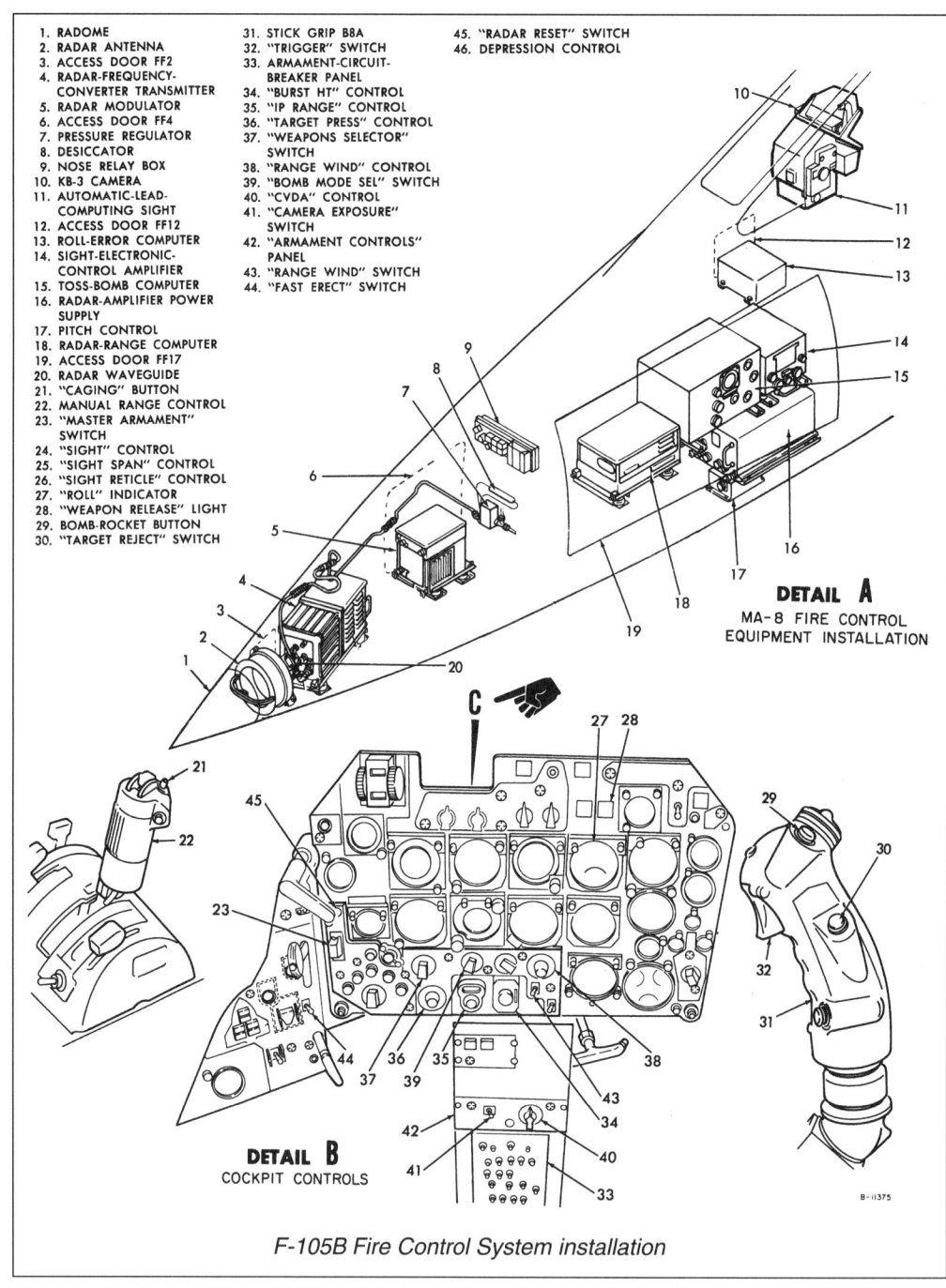

F-105B Fire Control System installation

At the time of its development, the MA-8 fire control system was the most complex ever fitted to a combat aircraft. But by the time the system was fully developed, its inherent limitations had already resulted in a decision to fit the all-weather F-105D with the even more advanced ASG-19. *(U.S. Air Force)*

November 1957 saw the approval of Thunderstick for the new F-105D and two-seat F-105E. However, serious budget restrictions a few months later brought a review of all ongoing projects and threatened to cancel some of the more advanced aspects of the new fire control system. Reducing the number of F-105s to be procured solved the majority of the funding problems, and the development of Thunderstick proceeded.

The most essential element of the new fire control system was the development of the North American Autonetics R-14A[5] NASSAR radar. At the time the Thunderstick project was approved, North American had already gained a great deal of experience with other versions of this radar, including the one planned for the production of F-107s that never materialized. North American was able to add a number of internal improvements to the radar planned for the F-107 and provide the new unit with the capability to perform contour mapping and terrain clearance. North American began testing the new radar aboard a B-26 during 1958.

By the beginning of 1959 a unit had been installed in an F-86K for further tests. The official Air Force evaluation of the radar came in April 1959 using the F-86K with "very satisfactory" results. Test pilots were impressed by the ground mapping, terrain clearance, and contour mapping modes of the radar. They found that the unit worked well when used with the APN-131 Doppler navigation set (an improved version of the APN-105). The tests did reveal some problems, however. The display was "ragged" due to oscillation of the angle of attack guide vanes in rough air, and the gyro drift was excessive, but neither problem was deemed serious enough to warrant modification, and the system was cleared for production and inclusion in the F-105D.

By the mid-1960s, tests of the ASG-19 at Eglin had been fairly successful, but many small deficiencies had been revealed, along with an overall reliability problem. Essentially the difficulties experienced at Eglin resulted from the complexity created by integrating the fire control system with the Doppler navigator, automatic flight control system, air data computer, and other components of the total F-105D weapons system. This was one of the first attempts at a truly integrated weapons system, and it showed the lack of experience. Still, by June 1960 the means to overcome these problems had been established and the changes were scheduled to be incorporated on the production line as part of Project Black Box. Aircraft already delivered would be modified in the field.

Concurrently, the capability of the system to employ nuclear weapons in the conventional dive- and toss-bombing modes was receiving less and less attention. By this time the Air Force was interested in the possibility of using drogue-retarded nuclear weapons that could be dropped from very low altitudes (≈300 feet). This was possible by retarding the fall of the bomb with a quick-opening parachute. Only minor changes and additions to the computer were required to permit such a bomb drop, and these were incorporated into the F-105 during 1961.

Although unsatisfactory from a reliability and maintainability perspective, the ASG-19 did prove to be a fairly capable fire control system over Vietnam. Given the technology of the day it was a tremendous advance over early nonintegrated weapons systems and gave valuable experience towards the design and manufacture of more advanced systems for the F-111.

Notes

1. *Development of Airborne Armament,* 1910–1961, Historical Division, Air Force Systems Command, Aeronautical Systems Division, Wright-Patterson AFB, Ohio, October 1961, pp III-430–451.
2. *The F-105: A Chronology (1951–1973),* History Office, Mobile Air Materiel Area, Brookley AFB, Alabama, September 1963, p 1.
3. At the end of December 1955, the Weapons Guidance laboratory had informed General Electric that it could no longer refer to the sight as the K-19 since that had been the designation of a purely experimental version of the sight built several years earlier. GE was, instead, to refer to the sight by a company part number. Regardless, most references continued to use K-19 in reference to any number of different versions of the sight.
4. U.S. nuclear weapons were known as TX-xx during their development, and became Mk xx when they entered the inventory.
5. This was a North American designation, not an Air Force one.

APPENDIX B

Acronyms

AAA	antiaircraft artillery
AB	Air Base
ADI	attitude director indicator
AEL	American Electronic Laboratories
AFB	Air Force Base
AFCS	automatic flight control system
AFRes	Air Force Reserve
AFT	aft crew station
Air Force	United States Air Force
AMI	airspeed mach indicator
ANG	Air National Guard
AP	advanced project (Republic designation)
APGC	Air Proving Ground Command
ARM	antiradiation missile
ATAF	Allied Tactical Air Force
ATC	Air Training Command
ATI	Advanced Technology, Incorporated
ATM	air turbine motor
AVVI	altitude vertical velocity indicator
az-el	azimuth-elevation
BASS	bistatic aided strike system
B/N	bombardier/navigator
BDA	bomb damage assessment
CCTS	Combat Crew Training Squadron

CCTW	Combat Crew Training Wing
CEP	circular error probable
CIN	communications, identification, and navigation
COMJAM	communication jamming
DECM	deception electronic countermeasures
DF	direction finding
ECM	electronic countermeasures
ELINT	electronic intelligence
EWO	electronic warfare officer
FAC	forward air controller
FEAF	Far East Air Forces
FFAR	folding fin aircraft rocket
FWD	forward crew station
FWS	Fighter Weapons Squadron
FWW	Fighter Weapons Wing
GCI	ground control intercept
GE	General Electric
GOR	general operating requirement
HIS	horizontal situation indicator
HP	high pressure
IFF	identification, friend or foe
ILS	instrument landing system
IOC	initial operational capability
KIAS	knots indicated air speed
LABS	low-altitude bombing system
LP	low pressure
LTV	Ling-Tempo-Vought
MER	multiple ejector rack
MHz	Megahertz
MiGCAP	MiG combat air patrol
MIT	Massachusetts Institute of Technology
MOAMA	Mobile (Alabama) Air Materiel Area
NAA	North American Aviation
NACA	National Advisory Committee on Aeronautics
NASA	National Aeronautics and Space Administration
NASARR	North American Search and Ranging Radar System
nav	navigation
O/R	operationally ready (rate)

OT&E	operational test and evaluation
P&W	Pratt & Whitney
PACAF	Pacific Air Forces
POL	petroleum, oil, lubricants
psi	pounds per square inch
QRC	quick-reaction capability
RAC	Republic Aviation Corporation
RAF	Royal Air Force (U.K.)
RAT	ram air turbine
RDT&E	research, development, test, and evaluation
RESCAP	rescue combat air patrol
RHAW	radar homing and warning
RN	radar-navigator
RTAFB	Royal Thai Air Force Base
RWR	radar warning receiver
SAC	Strategic Air Command
SAGE	Semi-Automatic Ground Environment
SAM	surface-to-air missile
SEA	Southeast Asia
SEE-SAMS	sense, exploit, and evade surface-to-air-missile system
SMAMA	Sacramento Air Materiel Area
SMART	supersonic military aircraft research track
SRW	Strategic Reconnaissance Wing
TAC	Tactical Air Command
TACAN	tactical air navigation
TAWC	Tactical Air Warfare Center
TBC	toss bomb computer
TCTO	time compliant technical order
TDY	temporary duty
TFG	Tactical Fighter Group
TFS	Tactical Fighter Squadron
TFTG	Tactical Fighter Training Group
TFW	Tactical Fighter Wing
TO	technical order
UHF	ultra-high frequency
USAF	United States Air Force
USAFE	United States Air Forces Europe
VAI	variable area inlet

Vdc	Volts direct current
VHF	very high frequency
WADC	Wright Air Development Center
WS	weapons system
WWS	Wild Weasel Squadron

APPENDIX C

Serial Numbers

Dash Number	Serial Number	Thru	No. of a/c	Comments
YF-105A-1-RE	54-0098	— 54-0099	2	
F-105B-1-RE	54-0100	— 54-0103	4	
F-105B-5-RE	54-0104		1	
RF-105B-1-RE	54-0105		1	to JF-105B
F-105B-5-RE	54-0106	— 54-0107	2	
RF-105B-1-RE	54-0108		1	to JF-105B
F-105B-5-RE	54-0109	— 54-0110	2	
F-105B-6-RE	54-0111		1	
RF-105B-2-RE	54-0112		1	to JF-105B
F-105B-10-RE	57-5776	— 57-5784	9	
F-105B-15-RE	57-5785	— 57-5802	18	
F-105B-20-RE	57-5803	— 57-5840	38	
F-105C-1-RE	54-0113	— 54-0117	5	canceled
F-105D-1-RE	58-1146	— 58-1148	3	
F-105D-5-RE	58-1149	— 58-1173	25	
F-105E-1-RE	58-1174	— 58-1175	2	canceled
F-105D-5-RE	59-1717	— 59-1757	41	
F-105D-6-RE	59-1758	— 59-1774	17	
	59-1817	— 59-1826	10	ordered as F-105E
F-105E-5-RE	59-1817	— 58-1842	26	canceled

189

Dash Number	Serial Number	Thru	No. of a/c	Comments
F-105D-6-RE	60-0409 — 60-0426		18	
F-105D-10-RE	60-0427 — 60-0535		109	
	60-5374 — 60-5385		12	
F-105D-15-RE	61-0041 — 61-0106		66	
F-105D-20-RE	61-0107 — 61-0161		55	
F-105D-25-RE	61-0162 — 61-0220		59	
	62-4217 — 62-4237		21	
F-105D-30-RE	62-4238 — 62-4276		39	
F-105D-31-RE	62-4277 — 62-4411		135	
F-105F-1-RE	62-4412 — 62-4447		36	ordered as F-105D
	63-8260 — 63-8366		107	
F-105Bs Modified to Thunderbird Configuration				
F-105B-10-RE	57-5782		1	Thunderbird
F-105B-15-RE	57-5787		1	Thunderbird
F-105B-15-RE	57-5790		1	Thunderbird
F-105B-15-RE	57-5793		1	Thunderbird (Slot)
F-105B-15-RE	57-5797		1	Thunderbird
F-105B-15-RE	57-5798		1	Thunderbird
F-105B-15-RE	57-5801		1	Thunderbird
F-105B-15-RE	57-5802		1	Thunderbird
F-105B-20-RE	57-5814		1	Thunderbird (Lead)
Total Thunderbird			9	
F-105Ds Modified with TCTO 1F-105D-700, -702, -703, and -704 (Thunderstick II)				
F-105D-10-RE	60-0455		1	T-Stick II Modification
F-105D-10-RE	60-0458		1	T-Stick II Modification
F-105D-10-RE	60-0464		1	T-Stick II Modification
F-105D-10-RE	60-0465		1	T-Stick II Modification
F-105D-10-RE	60-0471		1	T-Stick II Modification
F-105D-10-RE	60-0475		1	T-Stick II Modification
F-105D-10-RE	60-0480		1	T-Stick II Modification
F-105D-10-RE	60-0490		1	T-Stick II Modification
F-105D-10-RE	60-0493		1	T-Stick II Modification
F-105D-10-RE	60-0500		1	T-Stick II Modification
F-105D-10-RE	60-0513		1	T-Stick II Modification
F-105D-10-RE	60-0517		1	T-Stick II Modification

Appendix C—Serial Numbers

Dash Number	Serial Number	Thru	No. of a/c	Comments
F-105D-10-RE	60-0521		1	T-Stick II Modification
F-105D-10-RE	60-0527		1	T-Stick II Modification
F-105D-10-RE	60-0528		1	T-Stick II Modification
F-105D-10-RE	60-0533		1	T-Stick II Modification
F-105D-10-RE	60-5375		1	T-Stick II Modification
F-105D-10-RE	60-5376		1	T-Stick II Modification
F-105D-15-RE	61-0044		1	T-Stick II Modification
F-105D-15-RE	61-0047		1	T-Stick II Modification
F-105D-15-RE	61-0063		1	T-Stick II Modification
F-105D-15-RE	61-0064		1	T-Stick II Modification
F-105D-15-RE	61-0074		1	T-Stick II Modification
F-105D-15-RE	61-0075		1	T-Stick II Modification
F-105D-15-RE	61-0076		1	T-Stick II Modification
F-105D-15-RE	61-0080		1	T-Stick II Modification #1
F-105D-15-RE	61-0096		1	T-Stick II Modification
F-105D-15-RE	61-0100		1	T-Stick II Modification
F-105D-20-RE	61-0110		1	T-Stick II Modification
F-105D-20-RE	61-0161		1	T-Stick II Modification
F-105Fs Modified with TCTO 1F-105F-536 (Blind Bombing Mod)				
F-105F-1-RE	62-4419		1	
F-105F-1-RE	62-4424		1	Lost 1 May 1972
F-105F-1-RE	62-4428		1	
F-105F-1-RE	62-4429		1	Lost 15 May 1967
F-105F-1-RE	62-4446		1	
F-105F-1-RE	62-8274		1	
F-105F-1-RE	63-8263		1	
F-105F-1-RE	63-8269		1	Lost 12 May 1967
F-105F-1-RE	63-8274		1	
F-105F-1-RE	63-8275		1	
F-105F-1-RE	63-8276		1	
F-105F-1-RE	63-8277		1	
F-105F-1-RE	63-8278		1	
F-105F-1-RE	63-8281		1	Lost 21 February 1970
F-105F-1-RE	63-8285		1	
F-105F-1-RE	63-8293		1	Lost 18 February 1968

Dash Number	Serial Number	Thru	No. of a/c	Comments
F-105F-1-RE	63-8312		1	Lost 29 February 1968
F-105F-1-RE	63-8327		1	Written Off 12 March 1972
F-105F-1-RE	63-8346		1	Lost 4 October 1967
F-105F-1-RE	63-8353		1	Lost 15 July 1968
Total Ryan's Raiders			20	
F-105Fs Modified with Combat Martin				
F-105F-1-RE	62-4432		1	Combat Martin #2
F-105F-1-RE	62-4435		1	Combat Martin #6—Lost on 14 May 1969
F-105F-1-RE	62-4443		1	Combat Martin #9
F-105F-1-RE	62-4444		1	Combat Martin #5
F-105F-1-RE	63-8268		1	Combat Martin #7
F-105F-1-RE	63-8280		1	Combat Martin #10
F-105F-1-RE	63-8291		1	Combat Martin #8
F-105F-1-RE	63-8318		1	Combat Martin #3
F-105F-1-RE	63-8336		1	Combat Martin #4
F-105F-1-RE	63-8337		1	Combat Martin #1—Lost on 15 April 1968
Total Combat Martin			10	
F-105Fs Modified with BASS I/II				
F-105F-1-RE	63-8281		1	WWIII/RR
F-105F-1-RE	63-8285		1	WWIII/RR
F-105F-1-RE	63-8327		1	WWIII
F-105F-1-RE	63-8347		1	WWIII
Total BASS F-105F			4	
F-105Fs Modified with TCTO 1F-105F-522, -522C, -522D, -540, -547, -548, -550, and -1079 (Wild Weasel III) **TCTO 1F-105-1133 actually redesignated the aircraft as F-105G**				
F-105G-1-RE	62-4415		1	conversion canceled
F-105G-1-RE	62-4416		1	
F-105G-1-RE	62-4422		1	F-105G prototype #1
F-105G-1-RE	62-4423	—	62-4425	3	
F-105G-1-RE	62-4427	—	62-4428	2	
F-105G-1-RE	62-4432		1	
F-105G-1-RE	62-4434		1	F-105G prototype #2
F-105G-1-RE	62-4436		1	

Dash Number	Serial Number	Thru	No. of a/c	Comments
F-105G-1-RE	62-4438	—	62-4440	3
F-105G-1-RE	62-4442	—	62-4446	5
F-105G-1-RE	63-8265	—	63-8266	2
F-105G-1-RE	63-8274	—	63-8276	3
F-105G-1-RE	63-8278		1	
F-105G-1-RE	63-8281		1	conversion canceled
F-105G-1-RE	63-8284	—	63-8285	2
F-105G-1-RE	63-8291	—	63-8292	2
F-105G-1-RE	63-8296		1	
F-105G-1-RE	63-8300	—	63-8307	8
F-105G-1-RE	63-8311		1	
F-105G-1-RE	63-8313		1	
F-105G-1-RE	63-8316	—	63-8321	6
F-105G-1-RE	63-8326	—	63-8328	3
F-105G-1-RE	63-8332	—	63-8334	3
F-105G-1-RE	63-8336		1	
F-105G-1-RE	63-8339	—	63-8340	2
F-105G-1-RE	63-8342		1	
F-105G-1-RE	63-8345		1	
F-105G-1-RE	63-8347		1	
F-105G-1-RE	63-8350	—	63-8351	2
F-105G-1-RE	63-8355		1	
F-105G-1-RE	63-8359	—	63-8360	2
F-105G-1-RE	63-8363		1	
Total F-105G			**65**	
		Aircraft in Museums		
JF-105B-1-RE	54-0105	USAF History and Traditions Museum		
F-105B		Florida Military Aviation Museum		
F-105B-1-RE	54-0100	McClellan Aviation Museum		
F-105B-10-RE	57-5776	New Jersey ANG—108th ARW, Trenton		
F-105B-10-RE	57-5778	New England Air Museum		
F-105B-20-RE	57-5803	March Field Museum		
F-105B-20-RE	57-5837	Castle Air Museum		
F-105B-20-RE	57-5838	Wisconsin ANG—HQ		
F-105B-20-RE	57-5839	South Dakota Air and Space Museum		

		Aircraft in Museums
F-105D		United States Air Force Museum
F-105D-5-RE	58-1155	USAF Armament Museum
F-105D-5-RE	59-1738	Dyess Linear Air Park
F-105D-5-RE	59-1743	Hill AFB Museum
F-105D-10-RE	60-0482	USAF Academy
F-105D-10-RE	60-0508	Wings Over The Rockies Aviation & Space Museum
F-105D-10-RE	60-0535	Keesler AFB Air Park
F-105D-15-RE	61-0041	DC ANG—113rd FW, Andrews AFB
F-105D-15-RE	61-0050	Virginia ANG—192nd FG, Richmond
F-105D-15-RE	61-0056	Seymour Johnson AFB
F-105D-15-RE	61-0073	Air Power Park and Museum
F-105D-15-RE	61-0073	Langley A.F.B. Air Park
F-105D-15-RE	61-0086	Pima Air & Space Museum
F-105D-20-RE	61-0145	Holloman AFB
F-105D-20-RE	61-0146	Air Force Flight Test Center Museum
F-105D-20-RE	61-0159	Davis-Monthan AFB
F-105D-25-RE	61-0165	Pope AFB
F-105D-25-RE	61-0175	Sheppard AFB Air Park
F-105D-25-RE	61-0176	Maxwell AFB Air Park
F-105D-30-RE	62-4242	Vance AFB
F-105D-30-RE	62-4253	Reflections of Freedom Historical Air Park
F-105D-31-RE	62-4299	Travis Air Force Museum
F-105D-31-RE	62-4301	McClellan Aviation Museum
F-105D-31-RE	62-4328	Arnold Air Station
F-105D-31-RE	62-4383	March Field Museum
F-105D-31-RE	62-4387	USAF History and Traditions Museum
F-105F-1-RE	63-8366	McConnell AFB
F-105G-1-RE	62-4427	Pima Air & Space Museum
F-105G-1-RE	62-4440	Hill AFB Museum
F-105G-1-RE	62-4444	Empire State Aerosciences Museum
F-105G-1-RE	63-8276	Nellis AFB
F-105G-1-RE	63-8278	Silver Wings Aviation Museum
F-105G-1-RE	63-8285	Davis-Monthan AFB
F-105G-1-RE	63-8345	Georgia ANG—116th FW, Dobbins AFB
F-105G-1-RE		United States Air Force Museum
F-105G-1-RE		Museum of Aviation

APPENDIX D

F-105 Losses in Southeast Asia
Data from Center for Naval Analysis (CNA) Loss/Damage Database

F-105 Losses in Southeast Asi

	Combat Losses (direct or indirect)				
Loss #	Date	Aircraft	Serial No.	Unit	Cause
1	06 Jul 65	F-105D	62-4232	12 TFS/18 TFW	Automatic Weapons
2	14 Aug 64	F-105D	62-4371	36 TFS/41 AD	Unknown
3	13 Jan 65	F-105D	62-4296	18 TFW	AAA
4	02 Mar 65	F-105D	62-4214	67 TFS/18 TFW	AAA (37mm)
5	02 Mar 65	F-105D	62-4325	18 TFW	AAA
6	02 Mar 65	F-105D	62-4260	67 TFS/18 TFW	AAA
7	22 Mar 65	F-105D	62-4233	18 TFW	Small Arms
8	04 Apr 65	F-105D	62-4217	18 TFW	Unknown
9	04 Apr 65	F-105D	59-1764	355 TFW	MiG-17
10	04 Apr 65	F-105D	59-1754	355 TFW	MiG-17
11	05 Apr 65	F-105D	59-1742	355 TFW	AAA (37mm)
12	17 Apr 65	F-105D	61-0171	523 TFS/23 TFW	Combat Accident
13	07 May 65	F-105D	59-1718	354 TFS/355 TFW	Unknown
14	09 May 65	F-105D	62-4408	563 TFS/23 TFW	AAA
15	17 May 65	F-105D	62-4222	18 TFW	AAA (37mm)
16	18 May 65	F-105D	59-1731	563 TFS/35 TACG	AAA
17	23 May 65	F-105D	61-0054	354 TFS/355 TFW	Small Arms
18	31 May 65	F-105D	62-4381	354 TFS/355 TFW	Unknown
19	05 Jun 65	F-105D	61-0133	563 TFS/23 TFW	AAA
20	08 Jun 65	F-105D	62-4290	354 TFS/355 TFW	AAA
21	14 Jun 65	F-105D	62-4220	44 TFS/18 TFW	AAA
22	23 Jun 65	F-105D	62-4319	357 TFS/355 TFW	Automatic Weapons
23	21 Jul 65	F-105D	62-4257	563 TFS/23 TFW	AAA (37mm)
24	24 Jul 65	F-105D	62-4373	80 TFS/6441 TFW	AAA (37mm)
25	27 Jul 65	F-105D	62-4407	12 TFS/18 TFW	Automatic Weapons
26	27 Jul 65	F-105D	61-0113	563 TFS/23 TFW	AAA
27	27 Jul 65	F-105D	62-4252	12 TFS/18 TFW	AAA
28	27 Jul 65	F-105D	61-0177	355 CAMS/355 TFW	AAA
29	27 Jul 65	F-105D	62-4298	355 CAMS/355 TFW	Combat Mid Air
30	02 Aug 65	F-105D	62-4249	12 TFS/18 TFW	AAA (37mm)
31	03 Aug 65	F-105D	61-0098	355 CAMS/355 TFW	Automatic Weapons
32	10 Aug 65	F-105D	61-0184	18 TFW	AAA (37mm)
33	11 Aug 65	F-105D	61-0172	563 TFS/23 TFW	AAA (37mm)

Appendix D—F-105 Losses in Southeast Asia

Hit Location	Loss Location	Crew	Status
North Vietnam	North Vietnam	Capt. D. L. Williamson	MIA
Laos	Thailand	Unknown	Recovered Uninjured
Laos	Laos	Capt. A. Vallmer	Recovered Uninjured
North Vietnam	North Vietnam	1Lt. R. V. Baird	KIA
North Vietnam	Thailand	Capt. K. L. Spagnola	Recovered Uninjured
North Vietnam	Laos	Maj. G. W. Panas	Recovered Uninjured
North Vietnam	Loss at Sea	LtCol. R. Risner	Recovered Minor Injuries
North Vietnam	North Vietnam	Capt. C. S. Harris	POW—Returned
North Vietnam	North Vietnam	Capt. J. A. Magnusson, Jr.	KIA
North Vietnam	Loss at Sea	Maj. F. E. Bennett	KIA
North Vietnam	North Vietnam	Capt. T. Gay	Recovered Uninjured
North Vietnam	North Vietnam	Capt. S. A. Woodworth	KIA
North Vietnam	North Vietnam	Maj. R. E. Lambert	Recovered Minor Injuries
Laos	Laos	Capt. R. C. Wistrand	KIA
North Vietnam	North Vietnam	Capt. J. J. Tallaferro	Recovered Uninjured
Laos	Laos	Capt. D. L. Ardlicka	POW—MIA
North Vietnam	North Vietnam	Maj. R. F. Herman	Recovered Minor Injuries
North Vietnam	North Vietnam	1Lt. R. D. Peel	POW—Returned
Laos	Thailand	Capt. W. B. Kosko	Recovered Uninjured
North Vietnam	North Vietnam	Capt. H. Rademacher	Recovered Uninjured
North Vietnam	North Vietnam	Maj. Guarino	POW—Returned
North Vietnam	North Vietnam	Maj. R. W. Wilson	Recovered Uninjured
North Vietnam	North Vietnam	Capt. W. B. Kosko	MIA
Laos	Laos	Maj. W. L. McClelland	Recovered Uninjured
North Vietnam	North Vietnam	Capt. F. J. Tullo	Recovered Minor Injuries
North Vietnam	North Vietnam	Capt. K. D. Berg	POW—Returned
North Vietnam	North Vietnam	Capt. R. B. Purcell	POW—Returned
North Vietnam	Thailand	Capt. W. J. Barthelmas, Jr.	KIA
North Vietnam	Thailand	Maj. J. G. Farr	KIA
North Vietnam	North Vietnam	Capt. R. N. Daughtrey	POW—Returned
North Vietnam	North Vietnam	Maj. J. E. Bower	KIA
North Vietnam	Laos	Capt. M. J. Kelch	Recovered Minor Injuries
North Vietnam	North Vietnam	Capt. L. E. Wilson	Recovered Uninjured

F-105 Losses in Southeast Asia *(Continued)*

		Combat Losses (direct or indirect)			
Loss #	Date	Aircraft	Serial No.	Unit	Cause
34	22 Aug 65	F-105D	62-4235	36 TFS/6441 TFW	Unknown
35	25 Aug 65	F-105F	63-8282	67 TFS/18 TFW	Small Arms
36	29 Aug 65	F-105D	61-0193	67 TFS/18 TFW	AAA (37mm)
37	31 Aug 65	F-105D	61-0185	67 TFS/18 TFW	AAA (37mm)
38	02 Sep 65	F-105D	62-4389	36 TFS/6441 TFW	AAA (37mm)
39	06 Sep 65	F-105D	62-4337	67 TFS/18 TFW	Unknown
40	16 Sep 65	F-105D	61-0217	18 TFW	AAA (37mm)
41	16 Sep 65	F-105D	61-0189	18 TFW	AAA (37mm)
42	17 Sep 65	F-105D	62-4247	67 TFS/18 TFW	AAA (37mm)
43	20 Sep 65	F-105D	62-4238	67 TFS/18 TFW	Unknown
44	20 Sep 65	F-105D	61-0082	334 TFS/355 TFW	AAA
45	21 Sep 65	F-105D	61-0200	562 TFS/23 TFW	AAA (57mm)
46	30 Sep 65	F-105D	61-0119	334 TFS/355 TFW	SAM
47	05 Oct 65	F-105D	62-4376	36 TFS/6441 TFW	AAA
48	05 Oct 65	F-105D	62-4295	36 TFS/6441 TFW	AAA (37mm)
49	13 Oct 65	F-105D	61-0180	562 TFS/23 TFW	Unknown
50	14 Oct 65	F-105D	62-4305	36 TFS/6441 TFW	Unknown
51	14 Oct 65	F-105D	62-4333	36 TFS/6441 TFW	AAA (37mm)
52	22 Oct 65	F-105D	62-4350	36 TFS/6441 TFW	Automatic Weapons
53	03 Nov 65	F-105D	61-0163	562 TFS/23 TFW	Unknown
54	05 Nov 65	F-105D	62-4342	357 TFS/355 TFW	SAM
55	16 Nov 65	F-105D	62-4332	469 TFS/6234 TFW	SAM
56	18 Nov 65	F-105D	61-0062	469 TFS/6234 TFW	Unknown
57	28 Nov 65	F-105D	62-4285	355 TFW	Unknown
58	01 Dec 65	F-105D	61-0182	334 TFS/355 TFW	AAA (37mm)
59	15 Dec 65	F-105D	62-4363	334 TFS/355 TFW	AAA (37mm)
60	20 Dec 65	F-105D	61-0090	354 TFS/355 TFW	AAA (37mm)
61	21 Dec 65	F-105D	59-1823	421 TFS/6234 TFW	AAA (57mm)
62	11 Jan 66	F-105D	59-1736	334 TFS/355 TFW	Automatic Weapons
63	16 Jan 66	F-105D	59-1719	354 TFS/355 TFW	AAA (37mm)
64	31 Jan 66	F-105D	61-0210	469 TFS/388 TFW	AAA (37mm)
65	19 Feb 66	F-105D	62-4251	354 TFS/355 TFW	AAA (37mm)
66	26 Feb 66	F-105D	61-0215	421 TFS/388 TFW	Unknown

Hit Location	Loss Location	Crew	Status
North Vietnam	North Vietnam	Maj. D. A. Pogreba	Recovered Uninjured
North Vietnam	North Vietnam	Capt. W. D. Schlerman	POW—Returned
North Vietnam	North Vietnam	Maj. R. E. Byrne, Jr.	POW—Returned
North Vietnam	North Vietnam	Maj. W. H. Bollinger	Recovered Uninjured
North Vietnam	North Vietnam	Capt. J. O. Collins	POW—Returned
North Vietnam	Loss at Sea	Capt. J. T. Clark	Recovered Uninjured
North Vietnam	North Vietnam	LtCol. R. Risner	POW—Returned
North Vietnam	North Vietnam	Maj. R. J. Merritt	POW—Returned
North Vietnam	North Vietnam	1Lt. D. A. Klenda	KIA
North Vietnam	North Vietnam	Capt. E. L. Hawkins	KIA
North Vietnam	North Vietnam	Capt. W. E. Forby	POW—Returned
Laos	Laos	Capt. F. R. Greenwood	Recovered Minor Injuries
North Vietnam	North Vietnam	LtCol. M. J. Killian, Jr.	KIA
North Vietnam	North Vietnam	Capt. B. G. Seeber	POW—Returned
North Vietnam	North Vietnam	Maj. D. A. Pogreba	MIA
North Vietnam	North Vietnam	Maj. J. P. Randall	Recovered Minor Injuries
North Vietnam	North Vietnam	Capt. R. H. Schuler, Jr.	KIA
North Vietnam	North Vietnam	Capt. T. W. Sima	POW—Returned
North Vietnam	North Vietnam	Maj. F. V. Cherry	POW—Returned
North Vietnam	North Vietnam	Capt. D. P. Bowles	KIA
North Vietnam	North Vietnam	LtCol. G. C. McCleary	KIA
North Vietnam	Loss at Sea	Capt. D. G. Green	KIA
North Vietnam	North Vietnam	Capt. L. C. Mahaffey	Recovered Uninjured
North Vietnam	North Vietnam	Capt. J. A. Reynolds	POW—Returned
North Vietnam	North Vietnam	Capt. T. E. Reitmann	KIA
North Vietnam	North Vietnam	Capt. H. D. Dewitt	Recovered Minor Injuries
North Vietnam	North Vietnam	Capt. J. S. Ruffo	Recovered Minor Injuries
North Vietnam	North Vietnam	Capt. J. V. Sullivan	Recovered Uninjured
Laos	Laos	Capt. J. R. Stell	Recovered Uninjured
Laos	Laos	Capt. D. C. Wood	MIA
North Vietnam	North Vietnam	Capt. E. D. Hamilton	MIA
Laos	Laos	Capr. R. C. Green	Recovered Minor Injuries
Laos	Laos	Capt. C. G. Boyd	Recovered Minor Injuries

F-105 Losses in Southeast Asia *(Continued)*

		Combat Losses (direct or indirect)			
Loss #	Date	Aircraft	Serial No.	Unit	Cause
67	07 Mar 66	F-105D	62-4219	357 TFS/355 TFW	AAA
68	07 Mar 66	F-105D	62-4410	469 TFS/388 TFW	AAA (57mm)
69	16 Mar 66	F-105D	60-0411	333 TFS/355 TFW	AAA (37mm)
70	23 Mar 66	F-105D	60-0473	469 TFS/460 TFW	AAA (37mm)
71	23 Mar 66	F-105D	61-0178	357 TFS/355 TFW	AAA (37mm)
72	24 Mar 66	F-105D	61-0095	421 TFS/388 TFW	Unknown
73	19 Apr 66	F-105D	62-4330	333 TFS/355 TFW	Unknown
74	20 Apr 66	F-105D	60-0442	421 TFS/388 TFW	AAA (37mm)
75	22 Apr 66	F-105D	62-4409	421 TFS/388 TFW	AAA (37mm)
76	23 Apr 66	F-105D	61-0157	421 TFS/388 TFW	Unknown
77	23 Apr 66	F-105D	61-0048	421 TFS/388 TFW	AAA (37mm)
78	24 Apr 66	F-105D	62-4340	469 TFS/388 TFW	AAA (57mm)
79	24 Apr 66	F-105D	61-0051	469 TFS/388 TFW	SAM
80	29 Apr 66	F-105D	62-4304	333 TFS/355 TFW	AAA (85mm)
81	05 May 66	F-105D	61-0147	469 TFS/388 TFW	AAA (37mm)
82	06 May 66	F-105D	61-0179	388 TFW	AAA (57mm)
83	08 May 66	F-105D	62-4236	469 TFS/388 TFW	Small Arms
84	10 May 66	F-105D	61-0135	355 TFW	AAA (37mm)
85	10 May 66	F-105D	62-4255	469 TFS/388 TFW	Unknown
86	11 May 66	F-105D	62-4293	355 TFW	Unknown
87	15 May 66	F-105D	61-0174	421 TFS/388 TFW	Unknown
88	22 May 66	F-105D	58-1164	469 TFS/388 TFW	Unknown
89	25 May 66	F-105D	59-1746	354 TFS/355 TFW	Automatic Weapons
90	30 May 66	F-105D	61-0142	333 TFS/355 TFW	Unknown
91	31 May 66	F-105D	62-4386	354 TFS/355 TFW	AAA (85mm)
92	31 May 66	F-105D	61-0120	469 TFS/388 TFW	AAA (57mm)
93	01 Jun 66	F-105D	62-4394	357 TFS/355 TFW	Automatic Weapons
94	02 Jun 66	F-105D	59-1721	333 TFS/355 TFW	Automatic Weapons
95	03 Jun 66	F-105D	59-1171	469 TFS/388 TFW	AAA (37mm)
96	07 Jun 66	F-105D	60-0468	333 TFS/355 TFW	AAA (37mm)
97	08 Jun 66	F-105D	62-4273	334 TFS/355 TFW	Automatic Weapons
98	15 Jun 66	F-105D	62-4377	333 TFS/355 TFW	AAA (57mm)
99	21 Jun 66	F-105D	62-4358	388 TFW	AAA (85mm)

Hit Location	Loss Location	Crew	Status
North Vietnam	North Vietnam	Capt. H. V. Smith	MIA
North Vietnam	North Vietnam	Maj. J. L. Hutto	Recovered Minor Injuries
North Vietnam	North Vietnam	Maj. P. G. Underwood	MIA
North Vietnam	Laos	1Lt. K. D. Thomas	Recovered Minor Injuries
North Vietnam	Thailand	Maj. R. A. Hill	Recovered Minor Injuries
North Vietnam	North Vietnam	Capt. R. E. Bush	KIA
North Vietnam	North Vietnam	1Lt. L. A. Adams	KIA
North Vietnam	North Vietnam	Capt. J. B. Abernarthy	Recovered Uninjured
North Vietnam	North Vietnam	Capt. C. G. Boyd	POW—Returned
North Vietnam	North Vietnam	Capt. R. R. Dyczkowski	MIA
North Vietnam	North Vietnam	Maj. B. J. Goss	KIA
North Vietnam	North Vietnam	1Lt. J. D. Driscoll	POW—Returned
North Vietnam	North Vietnam	LtCol. W. E. Cooper	MIA
North Vietnam	North Vietnam	1Lt. D. W. Bruch, Jr.	KIA
North Vietnam	North Vietnam	1Lt. K. D. Thomas, Jr.	KIA
North Vietnam	North Vietnam	LtCol. J. L. Lamar	POW—Returned
North Vietnam	North Vietnam	1Lt. J. E. Ray	POW—Returned
North Vietnam	North Vietnam	Capt. M. H. Rahrt	Recovered Minor Injuries
North Vietnam	North Vietnam	Capt. J. E. Bailey	KIA
North Vietnam	North Vietnam	Capt. F. J. Feneley	KIA
North Vietnam	North Vietnam	Capt. R. C. Balcom	MIA
Laos	Laos	1Lt. R. M. Mackford	Recovered Uninjured
Laos	Laos	1Lt. R. G. Hunter	KIA
North Vietnam	North Vietnam	Capt. D. B. Hatcher	POW—Returned
North Vietnam	North Vietnam	1Lt. L. C. Ekman	Recovered Minor Injuries
North Vietnam	North Vietnam	Capt. M. W. Steen	KIA
North Vietnam	Laos	Capt. G. H. Peacock	Recovered Minor Injuries
North Vietnam	North Vietnam	Capt. J. D. Whipple	Recovered Uninjured
North Vietnam	Loss at Sea	Capt. R. D. Pielin	Recovered Minor Injuries
North Vietnam	Loss at Sea	Capt. J. F. Bayles	Recovered Uninjured
North Vietnam	Loss at Sea	Maj. J. C. Holley	Recovered Minor Injuries
North Vietnam	North Vietnam	1Lt. P. J. Kelly	Recovered Minor Injuries
North Vietnam	North Vietnam	1Lt. J. B. Sullivan III	MIA

F-105 Losses in Southeast Asia *(Continued)*

	Combat Losses (direct or indirect)				
Loss #	Date	Aircraft	Serial No.	Unit	Cause
100	29 Jun 66	F-105D	60-0460	333 TFS/355 TFW	AAA (85mm)
101	30 Jun 66	F-105D	62-4224	388 TFW	AAA (37mm)
102	01 Jul 66	F-105D	62-4354	388 TFW	AAA (37mm)
103	01 Jul 66	F-105D	59-1722	354 TFS/355 TFW	Unknown
104	04 Jul 66	F-105D	60-0486	355 TFW	AAA (37mm)
105	06 Jul 66	F-105D	62-4254	388 TFW	AAA (37mm)
106	06 Jul 66	F-105F (WW-III-1)	63-8286	13 TFS/388 TFW	AAA (57mm)
107	07 Jul 66	F-105D	59-1741	354 TFS/355 TFW	AAA (85mm)
108	11 Jul 66	F-105D	62-4282	355 TFW	AAA (85mm)
109	11 Jul 66	F-105D	61-0112	355 TFW	Automatic Weapons
110	11 Jul 66	F-105D	61-0121	355 TFW	MiG
111	15 Jul 66	F-105D	59-1761	388 TFW	AAA (37mm)
112	17 Jul 66	F-105D	58-1185	388 TFW	Unknown
113	19 Jul 66	F-105D	59-1755	354 TFS/355 TFW	MiG-17
114	19 Jul 66	F-105D	60-5382	354 TFS/355 TFW	AAA (57mm)
115	20 Jul 66	F-105D	62-4308	388 TFW	AAA (85mm)
116	20 Jul 66	F-105D	61-0116	355 TFW	Unknown
117	21 Jul 66	F-105D	62-4227	388 TFW	AAA (37mm)
118	23 Jul 66	F-105F	63-8338	354 TFS/355 TFW	SAM
119	25 Jul 66	F-105D	62-4271	469 TFS/388 TFW	Unknown
120	27 Jul 66	F-105D	61-0045	421 TFS/388 TFW	Unknown
121	01 Aug 66	F-105D	62-4380	388 TFW	AAA (57mm)
122	04 Aug 66	F-105D	61-0119	357 TFW/355 TFW	AAA (37mm)
123	06 Aug 66	F-105D	62-4315	421 TFS/388 TFW	AAA (37mm)
124	07 Aug 66	F-105D	62-4370	357 TFS/355 TFW	AAA (100mm)
125	07 Aug 66	F-105F	63-8358	354 TFS/355 TFW	SAM
126	07 Aug 66	F-105D	60-0499	333 TFS/355 TFW	SAM
127	07 Aug 66	F-105F	63-8361	354 TFS/355 TFW	AAA (85mm)
128	07 Aug 66	F-105D	61-0140	333 TFS/355 TFW	AAA (37mm)
129	08 Aug 66	F-105D	62-4327	354 TFS/355 TFW	AAA (37mm)
130	08 Aug 66	F-105D	61-0155	421 TFS/388 TFW	Unknown
131	08 Aug 66	F-105D	62-4343	354 TFS/355 TFW	AAA (37mm)
132	12 Aug 66	F-105D	62-4328	333 TFS/355 TFW	AAA (37mm)

Hit Location	Loss Location	Crew	Status
North Vietnam	North Vietnam	Capt. M. N. Jones	POW—Returned
North Vietnam	Laos	Capt. R. K. Hierste	Recovered Minor Injuries
North Vietnam	North Vietnam	1Lt. B. W. Campbell	POW—Returned
North Vietnam	North Vietnam	Capt. L. W. Shattuck	Recovered Uninjured
North Vietnam	Loss at Sea	1Lt. B. L. Minton	Recovered Uninjured
North Vietnam	Loss at Sea	Capt. E. L. Stanford	Recovered Uninjured
North Vietnam	North Vietnam	Maj. R. L. Hestle, Jr./Capt. C. E. Morgan	MIA/KIA
North Vietnam	North Vietnam	Capt. J. H. Jones	POW—Returned
North Vietnam	North Vietnam	Capt. L. W. Swattuck	POW—Returned
North Vietnam	Laos	Capt. R. H. Laney	Recovered Minor Injuries
North Vietnam	Laos	Maj. W. L. McClelland	Recovered Minor Injuries
North Vietnam	Loss at Sea	Capt. C. L. Hamby	Recovered Minor Injuries
North Vietnam	North Vietnam	1Lt. W. C. Spelius	Recovered Minor Injuries
North Vietnam	North Vietnam	1Lt. S. W. Diamond	MIA
North Vietnam	Laos	Capt. R. E. Steere	Recovered Minor Injuries
North Vietnam	North Vietnam	Capt. M. R. Lewis, Jr.	KIA
North Vietnam	North Vietnam	Col. W. H. Nelson	KIA
North Vietnam	North Vietnam	Capt. R. Tiffin	KIA
North Vietnam	North Vietnam	Maj. G. T. Pemberton/Maj. B. B. Newsom	POW—Died/POW—Died
North Vietnam	Thailand	Maj. F. C. Hiebert	Recovered Uninjured
North Vietnam	North Vietnam	Capt. J. R. Mitchell	Recovered Minor Injuries
North Vietnam	North Vietnam	Capt. K. W. North	POW—Returned
North Vietnam	Laos	1Lt. A. V. Rogers	Recovered Minor Injuries
North Vietnam	Loss at Sea	Capt. A. K. Rutherford	Recovered Minor Injuries
North Vietnam	North Vietnam	1Lt. M. L. Brazelton	POW—Returned
North Vietnam	Loss at Sea	Capt. E. Larsen/Capt. K. A. Gilroy	Recovered Minor Injuries
North Vietnam	North Vietnam	Capt. J. H. Wendell	POW—Returned
North Vietnam	North Vietnam	Capt. R. J. Sandvick/Capt. T. S. Pyle	POW—Returned
North Vietnam	North Vietnam	Maj. W. S. Gideon	POW—Returned
North Vietnam	North Vietnam	1Lt. F. R. Flom	POW—Returned
North Vietnam	North Vietnam	1Lt. J. R. Casper	Recovered Minor Injuries
North Vietnam	North Vietnam	Maj. J. H. Kasler	POW—Returned
North Vietnam	North Vietnam	1Lt. Nelems	POW—Returned

F-105 Losses in Southeast Asia *(Continued)*

	Combat Losses (direct or indirect)				
Loss #	Date	Aircraft	Serial No.	Unit	Cause
133	12 Aug 66	F-105D	61-0156	333 TFS/355 TFW	AAA (37mm)
134	14 Aug 66	F-105D	59-1763	333 TFS/355 TFW	AAA (37mm)
135	14 Aug 66	F-105D	61-0197	469 TFS/388 TFW	AAA (37mm)
136	14 Aug 66	F-105D	62-4266	354 TFS/355 TFW	AAA (37mm)
137	17 Aug 66	F-105F	63-8308	354 TFS/355 TFW	AAA
138	21 Aug 66	F-105D	59-1770	354 TFS/355 TFW	AAA (37mm)
139	29 Aug 66	F-105D	60-0523	354 TFS/355 TFW	Automatic Weapons
140	03 Sep 66	F-105D	62-4303	13 TFS/388 TFW	AAA (37mm)
141	04 Sep 66	F-105D	62-4369	357 TFS/355 TFW	AAA (57mm)
142	04 Sep 66	F-105D	61-0085	354 TFS/355 TFW	AAA (57mm)
143	05 Sep 66	F-105D	60-0495	354 TFS/355 TFW	Unknown
144	09 Sep 66	F-105D	62-4275	357 TFS/355 TFW	AAA (37mm)
145	12 Sep 66	F-105D	61-0201	469 TFS/388 TFW	AAA (85mm)
146	13 Sep 66	F-105D	62-4281	421 TFS/388 TFW	AAA (37mm)
147	14 Sep 66	F-105D	62-4306	421 TFS/388 TFW	AAA (57mm)
148	17 Sep 66	F-105D	61-0191	13 TFS/388 TFW	AAA (85mm)
149	17 Sep 66	F-105D	62-4280	469 TFS/388 TFW	AAA (37mm)
150	19 Sep 66	F-105D	62-4287	13 TFS/388 TFW	AAA (37mm)
151	21 Sep 66	F-105D	62-4321	357 TFS/355 TFW	AAA (85mm)
152	25 Sep 66	F-105D	62-4341	469 TFS/388 TFW	AAA (57mm)
153	26 Sep 66	F-105D	61-0186	13 TFS/388 TFW	AAA (37mm)
154	01 Oct 66	F-105D	60-0483	421 TFS/388 TFW	AAA (37mm)
155	14 Oct 66	F-105D	62-4391	354 TFS/355 TFW	Automatic Weapons
156	21 Oct 66	F-105D	61-0057	469 TFS/388 TFW	Unknown
157	27 Oct 66	F-105D	62-4396	333 TFS/355 TFW	Unknown
158	27 Oct 66	F-105D	60-0431	421 TFS/388 TFW	Automatic Weapons
159	02 Nov 66	F-105D	62-4379	421 TFS/388 TFW	Unknown
160	02 Nov 66	F-105D	60-0469	421 TFS/388 TFW	Unknown
161	04 Nov 66	F-105D	62-4366	469 TFS/388 TFW	AAA (37mm)
162	04 Nov 66	F-105F (WW-III-1)	63-8273	13 TFS/388 TFW	AAA (37mm)
163	06 Nov 66	F-105D	62-5487	355 TFW	AAA
164	06 Nov 66	F-105D	60-5374	355 TFW	AAA (37mm)
165	11 Nov 66	F-105D	62-4313	354 TFS/355 TFWA	Unknown

Appendix D—F-105 Losses in Southeast Asia

Hit Location	Loss Location	Crew	Status
North Vietnam	North Vietnam	Capt. D. J. Allinson	KIA
North Vietnam	North Vietnam	Capt. C. A. Eaton	MIA
North Vietnam	North Vietnam	Capt. C. E. Franklin	KIA
North Vietnam	North Vietnam	Capt. J. W. Brodak	POW—Returned
North Vietnam	North Vietnam	Maj. J. W. Brand/Maj. D. M. Singer	MIA/MIA
North Vietnam	Loss at Sea	Capt. N. L. Wells	Recovered Uninjured
North Vietnam	North Vietnam	Capt. N. L. Wells	POW—Returned
North Vietnam	Loss at Sea	Capt. E. R. Skowron	Recovered Minor Injuries
North Vietnam	North Vietnam	1Lt. R. G. Bliss	POW—Returned
North Vietnam	North Vietnam	1Lt. T. M. McNish	POW—Returned
North Vietnam	North Vietnam	Capt. T. D. Dobbs	Recovered Minor Injuries
North Vietnam	North Vietnam	Capt. J. C. Blevins	POW—Returned
North Vietnam	North Vietnam	Capt. R. F. Waggoner	POW—Returned
North Vietnam	North Vietnam	1Lt. K. V. Hallmark	Recovered Minor Injuries
North Vietnam	North Vietnam	1Lt. J. R. Casper	Recovered Minor Injuries
North Vietnam	South Vietnam	Capt. A. K. Rutherford	Recovered Minor Injuries
North Vietnam	North Vietnam	Capt. D. D. Leetum	KIA
North Vietnam	North Vietnam	Capt. D. G. Waltman	POW—Returned
North Vietnam	North Vietnam	Capt. G. L. Ammon	MIA
North Vietnam	North Vietnam	Capt. C. E. Cushman	KIA
North Vietnam	North Vietnam	Capt. A. T. Ballard	POW—Returned
North Vietnam	North Vietnam	Capt. C. G. Nix	POW—Returned
North Vietnam	North Vietnam	Maj. R. P. Taylor	Recovered Minor Injuries
Laos	Laos	Capt. C. P. Tofferi	KIA
North Vietnam	North Vietnam	Maj. D. A. Johnson	KIA
Laos	Laos	Maj. R. E. Kline	Recovered Uninjured
North Vietnam	Laos	Capt. R. F. Loken	Recovered Minor Injuries
North Vietnam	North Vietnam	Maj. R. E. Kline	MIA
North Vietnam	Laos	Capt. D. A. Elmer	Recovered Minor Injuries
North Vietnam	North Vietnam	Maj. R. E. Brinkman/Capt. V. A. Scungio	KIA
North Vietnam	North Vietnam	Capt. V. Vizcarra	Recovered Minor Injuries
North Vietnam	South Vietnam	Capt. W. G. Carey	Recovered Minor Injuries
North Vietnam	North Vietnam	Maj. A. S. Mearns	MIA

F-105 Losses in Southeast Asia *(Continued)*

			Combat Losses (direct or indirect)		
Loss #	Date	Aircraft	Serial No.	Unit	Cause
166	02 Dec 66	F-105D	59-1820	34 TFS/388 TFW	AAA (37mm)
167	05 Dec 66	F-105D	62-4331	421 TFS/388 TFW	MiG-17 (Gun)
168	06 Dec 66	F-105D	59-1725	354 TFS/355 TFW	AAA (37mm)
169	13 Dec 66	F-105D	61-0187	421 TFS/388 TFW	SAM
170	14 Dec 66	F-105D	60-0502	357 TFS/355 TFW	MiG-21 (AAM)
171	10 Jan 67	F-105D	62-4265	34 TFS/388 TFW	Unknown
172	15 Jan 67	F-105D	60-0440	333 TFS/355 TFW	Small Arms
173	21 Jan 67	F-105D	58-1156	421 TFS/388 TFW	AAA (85mm)
174	21 Jan 67	F-105D	62-4239	354 TFS/355 TFW	SAM
175	29 Jan 67	F-105F	62-4420	354 TFS/355 TFW	Unknown
176	30 Jan 67	F-105D	59-1768	469 TFS/388 TFW	Unknown
177	18 Feb 67	F-105F	63-8262	13 TFS/388 TFW	SAM
178	28 Feb 67	F-105D	59-1766	421 TFS/388 TFW	Unknown
179	04 Mar 67	F-105D	62-4274	357 TFS/355 TFW	AAA (37mm)
180	10 Mar 67	F-105F	63-8335	354 TFS/355 TFW	AAA (85mm)
181	11 Mar 67	F-105D	60-0443	333 TFS/355 TFW	AAA (57mm)
182	11 Mar 67	F-105D	62-4261	357 TFS/355 TFW	SAM
183	11 Mar 67	F-105D	60-0506	354 TFS/355 TFW	AAA (57mm)
184	15 Mar 67	F-105D	59-1825	357 TFS/355 TFW	Unknown
185	19 Mar 67	F-105D	61-0123	34 TFS/388 TFW	Unknown
186	26 Mar 67	F-105D	60-0516	469 TFS/388 TFW	Automatic Weapons
187	31 Mar 67	F-105D	59-1745	388 TFW	Automatic Weapons
188	02 Apr 67	F-105D	60-0426	13 TFS/388 TFW	AAA (37mm)
189	10 Apr 67	F-105D	62-4357	357 TFS/355 TFW	AAA (37mm)
190	14 Apr 67	F-105D	60-0447	357 TFS/355 TFW	AAA (37mm)
191	19 Apr 67	F-105F	63-8341	357 TFS/355 TFW	Unknown
192	25 Apr 67	F-105D	62-4294	354 TFS/355 TFW	AAA (85mm)
193	26 Apr 67	F-105D	58-1153	469 TFS/388 TFW	AAA
194	26 Apr 67	F-105F	63-8277	333 TFS/355 TFW	SAM
195	28 Apr 67	F-105D	58-1151	44 TFS/388 TFW	MiG-21 (Gun)
196	30 Apr 67	F-105D	61-0130	333 TFS/355 TFW	MiG-21 (AAM)
197	30 Apr 67	F-105D	59-1726	354 TFS/355 TFW	MiG-21 (AAM)
198	30 Apr 67	F-105F	63-4447	357 TFS/355 TFW	MiG-21 (AAM)

Hit Location	Loss Location	Crew	Status
North Vietnam	North Vietnam	Capt. M. L. Moorberg	KIA
North Vietnam	North Vietnam	Maj. B. N. Begley	MIA
North Vietnam	North Vietnam	LtCol. D. H. Asire	KIA
North Vietnam	North Vietnam	Capt. S. E. Waters	KIA
North Vietnam	North Vietnam	Capt. R. B. Cooley	Recovered Minor Injuries
Laos	Laos	Capt. J. P. Gauley	KIA
Laos	Thailand	Capt. G. L. Hawkins	Recovered Minor Injuries
North Vietnam	Loss at Sea	Capt. W. R. Wyatt	Recovered Major Injuries
North Vietnam	North Vietnam	LtCol. E. O. Conley	KIA
North Vietnam	North Vietnam	Maj. L. W. Biediger/1Lt. C. A. Silva	KIA/MIA
Laos	Thailand	Maj. W. E. Thurman	Recovered Uninjured
North Vietnam	North Vietnam	Capt. D. H. Duart/Capt. J. R. Jensen	POW—Returned
North Vietnam	North Vietnam	Capt. J. S. Walbridge	Recovered Minor Injuries
Laos	Laos	Maj. R. L. Carlock	KIA
North Vietnam	North Vietnam	Maj. D. Everson/Capt. J. D. Luna	POW—Returned
North Vietnam	North Vietnam	Capt. C. E. Greene, Jr.	POW—Returned
North Vietnam	North Vietnam	Capt. J. J. Karins, Jr.	KIA
North Vietnam	North Vietnam	Maj. J. E. Hiteshew	POW—Returned
North Vietnam	North Vietnam	LtCol. P. J. Frederick	MIA
North Vietnam	North Vietnam	LtCol. J. C. Austin	MIA
North Vietnam	South Vietnam	Maj. J. C. Spillers	Recovered Minor Injuries
North Vietnam	North Vietnam	Capt. H. J. Henningar	Recovered Minor Injuries
North Vietnam	North Vietnam	Capt. J. A. Dramesi	POW—Returned
North Vietnam	North Vietnam	Maj. J. F. O'Grady	MIA
North Vietnam	Laos	Maj. P. R. Craw	Recovered Major Injuries
North Vietnam	North Vietnam	Maj. T. M. Madison/Maj. T. J. Sterling	POW—Returned
North Vietnam	North Vietnam	1Lt. R. L. Weskamp	KIA
North Vietnam	North Vietnam	Capt. W. M. Meyer	KIA
North Vietnam	North Vietnam	Maj. J. F. Dudash/Capt. A. B. Meyers	KIA/POW—Returned
North Vietnam	North Vietnam	Capt. F. A. Caras	MIA
North Vietnam	North Vietnam	Capt. J. S. Abbott	POW—Returned
North Vietnam	North Vietnam	Capt. R. A. Abbott	POW—Returned
North Vietnam	North Vietnam	Maj. L. E. Thorsness/Capt. H. E. Johnson	POW—Returned

F-105 Losses in Southeast Asia *(Continued)*

	Combat Losses (direct or indirect)				
Loss #	Date	Aircraft	Serial No.	Unit	Cause
199	03 May 67	F-105D	62-4405	333 TFS/355 TFW	AAA (37mm)
200	05 May 67	F-105D	61-0198	357 TFS/355 TFW	AAA (37mm)
201	05 May 67	F-105D	62-4352	469 TFS/388 TFW	Unknown
202	05 May 67	F-105D	62-4401	469 TFS/388 TFW	AAA (85mm)
203	08 May 67	F-105D	61-0105	333 TFS/355 TFW	AAA (37mm)
204	12 May 67	F-105D	59-1728	357 TFS/355 TFW	Automatic Weapons
205	12 May 67	F-105F (RR)	63-8269	13 TFS/388 TFW	Unknown
206	14 May 67	F-105D	60-0421	13 TFS/388 TFW	SAM
207	15 May 67	F-105F (RR)	62-4429	13 TFS/388 TFW	AAA
208	27 May 67	F-105D	59-1723	333 TFS/355 TFW	SAM
209	02 Jun 67	F-105D	61-0190	34 TFS/388 TFW	AAA (85mm)
210	04 Jun 67	F-105D	61-0148	34 TFS/388 TFW	AAA (57mm)
211	15 Jun 67	F-105D	61-0213	388 TFW	AAA (37mm)
212	16 Jun 67	F-105D	60-0485	355 TFW	AAA
213	30 Jun 67	F-105D	62-4316	388 TFW	AAA (85mm)
214	02 Jul 67	F-105D	60-0494	44 TFS/388 TFW	Unknown
215	02 Jul 67	F-105D	60-0413	357 TFS/355 TFW	AAA (37mm)
216	05 Jul 67	F-105D	60-0454	357 TFS/355 TFW	AAA (57mm)
217	05 Jul 67	F-105D	61-0127	354 TFS/355 TFW	AAA (57mm)
218	05 Jul 67	F-105D	61-0042	357 TFS/355 TFW	AAA (85mm)
219	10 Jul 67	F-105D	60-0424	34 TFS/388 TFW	Automatic Weapons
220	13 Jul 67	F-105D	60-0450	357 TFS/355 TFW	AAA (37mm)
221	17 Jul 67	F-105D	59-1748	333 TFS/355 TFW	AAA (85mm)
222	19 Jul 67	F-105D	60-0441	357 TFS/355 TFW	AAA (37mm)
223	28 Jul 67	F-105D	62-4334	34 TFS/388 TFW	AAA (57mm)
224	29 Jul 67	F-105D	58-1163	357 TFS/355 TFW	Unknown
225	03 Aug 67	F-105D	58-1154	13 TFS/388 TFW	AAA (85mm)
226	17 Aug 67	F-105D	62-4378	44 TFS/388 TFW	AAA (37mm)
227	21 Aug 67	F-105D	60-0437	354 TFS/355 TFW	AAA
228	21 Aug 67	F-105D	59-1720	354 TFS/355 TFW	AAA
229	23 Aug 67	F-105D	59-1752	357 TFS/355 TFW	AAA (85mm)
230	24 Aug 67	F-105D	62-4268	357 TFS/355 TFW	AAA (37mm)
231	02 Sep 67	F-105D	62-4338	333 TFS/355 TFW	Unknown

Hit Location	Loss Location	Crew	Status
North Vietnam	North Vietnam	Maj. C. C. Vasiliadis	Recovered Minor Injuries
North Vietnam	North Vietnam	1Lt. J. R. Shively	POW—Returned
North Vietnam	North Vietnam	LtCol. G. A. Larson	POW—Returned
North Vietnam	North Vietnam	LtCol. J. L. Highes	POW—Returned
North Vietnam	North Vietnam	Capt. M. K. McCuistion	POW—Returned
North Vietnam	North Vietnam	Capt. E. W. Grenzebach, Jr.	MIA
North Vietnam	North Vietnam	Capt. R. A. Stewart/Capt. P. P. Pitman	KIA/MIA
North Vietnam	North Vietnam	Maj. G. R. Wilson	Recovered Minor Injuries
North Vietnam	North Vietnam	Maj. B. M. Pollard/Capt. D. L. Heiliger	POW—Returned
North Vietnam	North Vietnam	Capt. G. B. Blackwood	MIA
North Vietnam	North Vietnam	Maj. D. L. Smith	POW—Returned
North Vietnam	Loss at Sea	Capt. C. J. Kough, Jr.	Recovered Minor Injuries
North Vietnam	Loss at Sea	Capt. J. W. Swanson, Jr.	MIA
North Vietnam	North Vietnam	LtCol. W. L. Janssen	Recovered Minor Injuries
North Vietnam	Laos	Maj. R. L. Kuster	Recovered Minor Injuries
North Vietnam	North Vietnam	Capt. D. M. Pichard	Recovered Minor Injuries
North Vietnam	North Vietnam	Maj. R. E. Stone	Recovered Minor Injuries
North Vietnam	North Vietnam	Capt. W. V. Frederick	MIA
North Vietnam	North Vietnam	Maj. D. W. Waddell	POW—Returned
North Vietnam	North Vietnam	Maj. W. K. Dodge	POW—Died
North Vietnam	Laos	Maj. M. E. Seaver, Jr.	Recovered Uninjured
North Vietnam	Thailand	Maj. C. D. Osborne	Recovered Minor Injuries
North Vietnam	North Vietnam	Maj. H. C. Copeland	POW—Returned
North Vietnam	Loss at Sea	Capt. W. N. Johnson	Recovered Uninjured
North Vietnam	Laos	1Lt. K. W. Richter	KIA (Died after rescue)
North Vietnam	North Vietnam	1Lt. J. B. West	Recovered Minor Injuries
North Vietnam	North Vietnam	Capt. W. G. Newcomb	POW—Returned
North Vietnam	Loss at Sea	Maj. A. C. Vollmer	Recovered Major Injuries
North Vietnam	North Vietnam	1Lt. L. K. Powell	KIA
North Vietnam	North Vietnam	Capt. M. L. Morrill	MIA
North Vietnam	North Vietnam	Maj. E. C. Baker	POW—Returned
North Vietnam	North Vietnam	Capt. J. C. Hess	POW—Returned
North Vietnam	North Vietnam	Maj. W. G. Bennett	KIA

F-105 Losses in Southeast Asia *(Continued)*

	Combat Losses (direct or indirect)				
Loss #	Date	Aircraft	Serial No.	Unit	Cause
232	03 Sep 67	F-105D	61-0078	469 TFS/388 TFW	AAA (37mm)
233	23 Sep 67	F-105D	59-1749	469 TFS/388 TFW	AAA (85mm)
234	03 Oct 67	F-105D	59-1727	469 TFS/388 TFW	SAM
235	04 Oct 67	F-105F (RR)	63-8346	13 TFS/388 TFW	Unknown
236	05 Oct 67	F-105D	58-1169	13 TFS/388 TFW	AAA (85mm)
237	07 Oct 67	F-105F	63-8330	13 TFS/388 TFW	MiG-21 (AAM)
238	07 Oct 67	F-105D	60-0444	34 TFS/388 TFW	AAA (85mm)
239	09 Oct 67	F-105D	60-0434	34 TFS/388 TFW	MiG-21 (AAM)
240	17 Oct 67	F-105D	61-0205	34 TFS/388 TFW	AAA (37mm)
241	17 Oct 67	F-105D	62-4326	34 TFS/388 TFW	AAA (85mm)
242	17 Oct 67	F-105D	60-0425	34 TFS/388 TFW	AAA (85mm)
243	24 Oct 67	F-105D	62-4262	354 TFS/355 TFW	AAA (85mm)
244	25 Oct 67	F-105D	59-1735	333 TFS/355 TFW	AAA (57mm)
245	25 Oct 67	F-105D	58-1168	354 TFS/355 TFW	AAA (37mm)
246	27 Oct 67	F-105D	61-0126	469 TFS/388 TFW	AAA (37mm)
247	27 Oct 67	F-105D	62-4231	469 TFS/388 TFW	SAM
248	27 Oct 67	F-105D	61-0122	357 TFS/355 TFW	SAM
249	28 Oct 67	F-105D	61-0169	357 TFS/355 TFW	AAA (37mm)
250	05 Nov 67	F-105D	61-0173	333 TFS/355 TFW	Unknown
251	05 Nov 67	F-105F	62-4430	357 TFS/355 TFW	AAA (37mm)
252	06 Nov 67	F-105D	62-4286	469 TFS/388 TFW	SAM
253	07 Nov 67	F-105D	60-0430	469 TFS/388 TFW	AAA (37mm)
254	08 Nov 67	F-105D	61-0094	354 TFS/355 TFW	AAA (37mm)
255	17 Nov 67	F-105D	62-4258	354 TFS/355 TFW	SAM
256	18 Nov 67	F-105D	62-4221	44 TFS/388 TFW	SAM
257	18 Nov 67	F-105D	60-0497	469 TFS/388 TFW	MiG-21 (AAM)
258	18 Nov 67	F-105D	62-4283	469 TFS/388 TFW	SAM
259	18 Nov 67	F-105F	63-8295	44 TFS/388 TFW	MiG-21 (AAM)
260	19 Nov 67	F-105D	61-0208	469 TFS/388 TFW	SAM
261	19 Nov 67	F-105F	63-8349	333 TFS/355 TFW	SAM
262	19 Nov 67	F-105D	58-1170	34 TFS/388 TFW	SAM
263	20 Nov 67	F-105D	61-0124	460 TFS/388 TFW	MiG-21 (AAM)
264	05 Dec 67	F-105D	59-1758	333 TFS/355 TFW	AAA (37mm)

Hit Location	Loss Location	Crew	Status
North Vietnam	North Vietnam	Capt. H. W. Moore, Jr.	MIA
North Vietnam	Thailand	Maj. D. S. Aunapu	Recovered Minor Injuries
North Vietnam	North Vietnam	Maj. R. W. Barnett	POW—Returned
North Vietnam	North Vietnam	Maj. M. L. McDaniel, Jr./Capt. W. A. Lillund	MIA/MIA
North Vietnam	North Vietnam	Capt. K. W. Trautman	POW—Returned
North Vietnam	Loss at Sea	Capt. J. C. Howard/Capt. G. L. Shamblee	Recovered Minor Injuries
North Vietnam	North Vietnam	Maj. W. E. Fullam	KIA
North Vietnam	North Vietnam	Maj. J. A. Clements	POW—Returned
North Vietnam	North Vietnam	Capt. A. C. Andrews	POW—Returned
North Vietnam	North Vietnam	Capt. D. E. O'Dell	POW—Returned
North Vietnam	North Vietnam	Maj. D. E. Sullivan	POW—Returned
North Vietnam	North Vietnam	Capt. M. D. Scott	Recovered Uninjured
North Vietnam	North Vietnam	Capt. R. A. Horinek	POW—Returned
North Vietnam	North Vietnam	Maj. R. E. Smith, Jr.	POW—Returned
North Vietnam	North Vietnam	Capt. R. E. Temperley	POW—Returned
North Vietnam	North Vietnam	Col. J. P. Flynn	POW—Returned
North Vietnam	North Vietnam	Maj. R. L. Stirm	POW—Returned
North Vietnam	North Vietnam	LtCol. T. H. Kirk, Jr.	POW—Returned
North Vietnam	North Vietnam	Capt. B. R. Sparks	Recovered Uninjured
North Vietnam	North Vietnam	Maj. R. A. Dutton/Capt. E. G. Cobell	POW—Returned/POW—Died
North Vietnam	North Vietnam	Maj. R. W. Hagerman	KIA
North Vietnam	North Vietnam	Maj. W. C. Diehl, Jr.	POW—Died
North Vietnam	North Vietnam	Capt. L. G. Evert	MIA
North Vietnam	North Vietnam	Maj. C. E. Cappelli	KIA
North Vietnam	North Vietnam	Col. E. B. Burdett	POW—Died
North Vietnam	North Vietnam	LtCol. W. N. Reed	Recovered Uninjured
North Vietnam	North Vietnam	Maj. L. J. Hauer	KIA
North Vietnam	North Vietnam	Maj. O. M. Dardeau, Jr./Capt. E. W. Lenhoff	KIA/KIA
North Vietnam	North Vietnam	Capt. H. H. Klinck	KIA
North Vietnam	Laos	Maj. G. C. Gustafson/Capt. R. F. Brownlee	Recovered Minor Injuries/Uninjured
North Vietnam	North Vietnam	Maj. R. W. Vissotzky	POW—Returned
North Vietnam	North Vietnam	Capt. W. W. Butler	POW—Returned
Laos	Laos	Maj. D. M. Russell	KIA

F-105 Losses in Southeast Asia *(Continued)*

			Combat Losses (direct or indirect)		
Loss #	Date	Aircraft	Serial No.	Unit	Cause
265	14 Dec 67	F-105D	59-1750	469 TFS/388 TFW	AAA (85mm)
266	17 Dec 67	F-105D	60-0422	469 TFS/388 TFW	MiG-21 (AAM)
267	03 Jan 68	F-105D	58-1157	469 TFS/388 TFW	MiG-21 (AAM)
268	05 Jan 68	F-105D	61-0068	469 TFS/388 TFW	AAA (85mm)
269	05 Jan 68	F-105F	63-8356	357 TFS/355 TFW	Mig-17 (Gun)
270	14 Jan 68	F-105D	60-0489	469 TFS/388 TFW	MiG-21 (AAM)
271	04 Feb 68	F-105D	60-5384	34 TFS/388 TFW	MiG-21 (AAM)
272	14 Feb 68	F-105D	60-0418	34 TFS/388 TFW	SAM
273	18 Feb 68	F-105F	63-8293	44 TFS/388 TFW	AAA (37mm)
274	26 Feb 68	F-105D	62-4385	354 TFS/355 TFW	Unknown
275	29 Feb 68	F-105F (RR/WW)	63-8312	44 TFS/388 TFW	SAM
276	06 Mar 68	F-105D	62-4336	333 TFS/355 TFW	Unknown
277	17 Mar 68	F-105D	61-0162	469 TFS/388 TFW	Automatic Weapons
278	15 Apr 68	F-105F (CM)	63-8337	357 TFS/355 TFW	AAA (37mm)
279	15 Apr 68	F-105D	61-0207	34 TFS/388 TFW	AAA (37mm)
280	10 May 68	F-105D	60-0415	357 TFS/355 TFW	AAA (50-cal)
281	28 May 68	F-105D	61-0194	34 TFS/388 TFW	Automatic Weapons
282	30 May 68	F-105D	60-0511	469 TFS/388 TFW	Automatic Weapons
283	31 May 68	F-105D	60-0409	469 TFS/388 TFW	Unknown
284	08 Jun 68	F-105D	61-0055	34 TFS/388 TFW	AAA (37mm)
285	23 Jun 68	F-105D	59-1765	333 TFS/355 TFW	AAA (37mm)
286	01 Jul 68	F-105D	61-0118	333 TFS/355 TFW	Automatic Weapons
287	13 Jul 68	F-105D	60-0453	34 TFS/388 TFW	AAA
288	14 Jul 68	F-105D	62-4367	333 TFS/355 TFW	AAA (37mm)
289	15 Jul 68	F-105F	63-8353	44 TFS/388 TFW	AAA (37mm)
290	09 Aug 68	F-105D	62-4292	357 TFS/355 TFW	AAA (37mm)
291	01 Sep 68	F-105D	60-0512	34 TFS/388 TFW	AAA
292	12 Sep 68	F-105D	59-1762	357 TFS/355 TFW	Unknown
293	14 Sep 68	F-105D	60-0522	357 TFS/355 TFW	AAA (37mm)
294	19 Sep 68	F-105D	60-0428	469 TFS/388 TFW	AAA (37mm)
295	30 Sep 68	F-105F	63-8317	333 TFS/355 TFW	AAA (37mm)
296	27 Oct 68	F-105D	62-4264	34 TFS/388 TFW	AAA (37mm)
297	06 Dec 68	F-105D	61-0053	357 TFS/355 TFW	AAA (57mm)

Hit Location	Loss Location	Crew	Status
North Vietnam	North Vietnam	Capt. J. E. Sehorn	POW—Returned
North Vietnam	North Vietnam	1Lt. J. T. Ellis	POW—Returned
North Vietnam	North Vietnam	Col. J. E. Bean	POW—Returned
North Vietnam	North Vietnam	Capt. J. E. Jones	MIA
North Vietnam	North Vietnam	Maj. J. C. Hartney/Capt. S. Fantle III	KIA/KIA
North Vietnam	North Vietnam	Maj. S. H. Horne	KIA
North Vietnam	North Vietnam	Capt. C. W. Lasiter	POW—Returned
North Vietnam	North Vietnam	Capt. R. M. Elliot	MIA
Laos	Thailand	Maj. M. S. Muskat/Capt. K. Stouder	Recovered Minor Injuries
Laos	Laos	Capt. G. Basel	Recovered Major Injuries
North Vietnam	North Vietnam	Maj. C. J. Fitton, Jr./Capt. C. S. Harris	KIA/KIA
Laos	Thailand	Capt. F. E. Peck	Recovered Major Injuries
Laos	Laos	Capt. T. T. Hensley	KIA
North Vietnam	Loss at Sea	Col. D. W. Winn	Recovered Injured
North Vietnam	North Vietnam	Maj. J. H. Metz	POW—Died
South Vietnam	South Vietnam	Maj. D. B. Coons	Recovered Minor Injuries
North Vietnam	Laos	Maj. R. E. Ingvalson	POW—Returned
Laos	Laos	Col. N. P. Phillips	Recovered Minor Injuries
North Vietnam	Loss at Sea	Maj. E. P. Beresik	KIA
North Vietnam	North Vietnam	Maj. C. B. Light	Recovered
North Vietnam	Loss at Sea	Maj. J. W. Aldep	Recovered
North Vietnam	North Vietnam	LtCol. J. Modica, Jr.	Recovered Minor Injuries
North Vietnam	Thailand	1Lt. G. R. Confer	Recovered Minor Injuries
North Vietnam	North Vietnam	Maj. R. K. Hanna	Recovered Injured
North Vietnam	North Vietnam	Maj. G. D. James/Capt. L. E. Martin	POW—Returned/KIA
North Vietnam	North Vietnam	Col. D. W. Wind	POW—Returned
Laos	Laos	Capt. D. K. Thaete	Recovered Injured
North Vietnam	North Vietnam	Maj. S. C. Maxwell	KIA
North Vietnam	Laos	Capt. D. M. Thibble	Recovered Uninjured
North Vietnam	North Vietnam	Capt. E. R. Capling	KIA
North Vietnam	North Vietnam	Capt. C. W. Fieszel/Maj. H. H. Smith	MIA/MIA
North Vietnam	North Vietnam	1Lt. R. C. Edmunds, Jr.	KIA
Laos	Laos	Capt. R. M. Walker	Recovered Uninjured

F-105 Losses in Southeast Asia (Continued)

		Combat Losses (direct or indirect)			
Loss #	Date	Aircraft	Serial No.	Unit	Cause
298	08 Dec 68	F-105D	61-0150	354 TFS/355 TFW	Unknown
299	21 Dec 68	F-105D	61-0089	354 TFS/355 TFW	Unknown
300	24 Dec 68	F-105D	62-4234	354 TFS/355 TFW	Unknown
301	11 Jan 69	F-105D	61-0072	354 TFS/355 TFW	AAA (37mm)
302	11 Feb 69	F-105D	62-4256	34 TFS/388 TFW	Automatic Weapons
303	12 Feb 69	F-105D	60-0417	333 TFS/355 TFW	Automatic Weapons
304	18 Feb 69	F-105D	60-0505	34 TFS/388 TFW	AAA (37mm)
305	02 Mar 69	F-105D	61-0109	333 TFS/355 TFW	Automatic Weapons
306	17 Mar 69	F-105D	61-0104	34 TFS/388 TFW	Automatic Weapons
307	29 Mar 69	F-105D	62-4270	34 TFS/388 TFW	AAA (37mm)
308	03 Apr 69	F-105D	62-4269	34 TFS/388 TFW	Automatic Weapons
309	14 May 69	F-105F (CM)	62-4435	354 TFS/355 TFW	Automatic Weapons
310	14 Jun 69	F-105D	60-5381	354 TFS/355 TFW	Automatic Weapons
311	16 Jun 69	F-105D	60-0530	354 TFS/355 TFW	Small Arms
312	15 Jul 69	F-105D	60-0518	44 TFS/388 TFW	Unknown
313	25 Aug 69	F-105D	59-1818	357 TFS/355 TFW	Unknown
314	04 Nov 69	F-105D	59-1734	357 TFS/355 TFW	Unknown
315	24 Nov 69	F-105D	61-0060	357 TFS/355 TFW	Unknown
316	08 Dec 69	F-105F	63-8352	357 TFS/355 TFW	AAA (37mm)
317	27 Jan 70	F-105D	59-1772	333 TFS/355 TFW	AAA (37mm)
318	28 Jan 70	F-105G	63-8329	44 TFS/388 TFW	AAA
319	21 Feb 70	F-105G	63-8281	354 TFS/355 TFW	Small Arms
320	16 Mar 70	F-105D	62-4230	354 TFS/355 TFW	Automatic Weapons
321	20 Apr 70	F-105D	60-0451	333 TFS/355 TFW	Automatic Weapons
322	23 Sep 70	F-105D	61-0153	355 TFW	AAA (37mm)
323	20 Nov 70	F-105G	62-4436	388 TFW	Unknown
324	10 Dec 71	F-105G	63-8326	17 WWS/388 TFW	SAM
325	17 Feb 72	F-105G	63-8333	17 WWS/388 TFW	SAM
326	15 Apr 72	F-105G	63-8342	17 WWS/388 TFW	SAM
327	11 May 72	F-105G	82-4424	17 WWS/388 TFW	Mig-21 (AAM)
328	29 Jul 72	F-105G	62-4443	17 WWS/388 TFW	Own Missile
329	17 Sep 72	F-105G	63-8360	17 WWS/388 TFW	SAM
330	29 Sep 72	F-105G	63-8302	17 TFS/388 TFW	SAM
331	16 Nov 72	F-105G	63-8359	561 TFS/388 TFW	SAM

Hit Location	Loss Location	Crew	Status
Laos	Laos	1Lt. R. A. Rex	KIA
Laos	Laos	Capt. R. K. Allee	MIA
Laos	Laos	Maj. C. R. Brownlee	MIA
Laos	Laos	Maj. W. M. Thompson	Recovered Injured
Laos	Laos	1Lt. R. J. Zukowski	KIA
Laos	Thailand	Maj. V. Colasuonno	KIA
Laos	North Vietnam	Capt. J. M. Brucher	KIA
Laos	Laos	Maj. C. C. Bogiages, Jr.	MIA
Laos	Laos	1Lt. D. T. Dinan III	KIA
Laos	Laos	1Lt. R. A. Stafford	Recovered Minor Injuries
Laos	Laos	Maj. P. B. Christianson	KIA
Laos	Thailand	Maj. A. M. Yahanda	Recovered Minor Injuries
Laos	Laos	Maj. H. Kahler	MIA
Laos	Laos	1Lt. J. L. Devoss	Recovered Injured
Laos	Thailand	Maj. R. E. Kennedy	Recovered Minor Injuries
Laos	Laos	Maj. S. R. Sanders	KIA
Laos	Laos	Capt. L. J. Hanley	MIA
Laos	Laos	Capt. J. B. White	KIA
Laos	Thailand	Maj. C. R. Dice/1Lt. B. N. Cox	KIA/Recovered Minor Injuries
Laos	Laos	Maj. D. W. Livingston	Recovered Minor Injuries
North Vietnam	North Vietnam	Capt. R. J. Mallon/Capt. R. J. Panek, Sr.	KIA/KIA
Laos	Thailand	Maj. G. B. Hurst/Capt. C. S. Bevan	Recovered Minor Injuries
Laos	Thailand	Maj. W. J. Wycoff	Recovered Minor Injuries
Laos	Laos	Capt. D. F. Mahan	KIA
Laos	Laos	Capt. J. W. Newhouse	Recovered Minor Injuries
Laos	Laos	Maj. D. W. Kilgus/Capt. C. T. Lowry	Recovered Minor Injuries
Laos	Laos	LtCol. S. W. McIntire/Maj. R. E. Belli	KIA/Recovered Minor Injuries
North Vietnam	Loss at Sea	Capt. J. D. Cutter/Capt. K. J. Fraser	POW—Returned
North Vietnam	North Vietnam	Capt. A. P. Mateja/Capt. O. C. Jones, Jr.	MIA/MIA
North Vietnam	North Vietnam	Maj. W. H. Talley/Maj. J. P. Padgett	POW—Returned
North Vietnam	Loss at Sea	Maj. T. J. Coady/H. F. Murphy	Recovered
North Vietnam	Loss at Sea	Capt. T. O. Zorn/1Lt. M. S. Turose	KIA/KIA
North Vietnam	North Vietnam	LtCol. J. W. Oneil/Capt. M. J. Bosiljevac	POW—Returned/POW—Died
North Vietnam	North Vietnam	Maier/Thaete	Unknown

F-105 Losses in Southeast Asia *(Continued)*

	Losses to Other than Combat (accidents, etc.)			
Loss #	Date	Aircraft	Serial No.	Unit
332	12 May 65	F-105D	61-0125	563 TFS/23 TFW
333	15 May 65	F-105D	62-4374	6441 TFW
334	03 Jul 65	F-105D	62-4398	563 TFS/23 TFW
335	30 Aug 65	F-105D	62-4355	
336	06 Sep 65	F-105D	62-4400	562 TFS/23 TFW
337	28 Sep 65	F-105D	62-4404	23 TFW
338	12 Nov 65	F-105D	62-4218	562 TFS/23 TFW
339	08 Dec 65	F-105D	62-4302	6234 TFW
340	20 Jan 66	F-105D	59-1717	333 TFS/355 TFW
341	20 Jan 66	F-105D	62-4324	333 TFS/355 TFW
342	22 Feb 66	F-105D	62-4388	357 TFS/355 TFW
343	27 Feb 66	F-105D	62-4362	355 OMS/355 TFW
344	02 Mar 66	F-105D	59-1724	355 OMS/355 TFW
345	15 Apr 66	F-105D	58-1158	469 TFS/388 TFW
346	13 May 66	F-105D	60-0427	421 TFS/388 TFW
347	02 Jun 66	F-105D	61-0160	421 TFS/388 TFW
348	14 Jun 66	F-105D	60-0429	421 TFS/388 TFW
349	18 Jul 66	F-105D	62-4312	421 TFS/388 TFW
350	10 Oct 66	F-105D	62-4300	469 TFS/388 TFW
351	10 Nov 66	F-105D	62-4288	469 TFS/388 TFW
352	22 Nov 66	F-105D	58-1161	469 TFS/388 TFW
353	10 Dec 66	F-105D	58-1160	357 TFS/355 TFW
354	17 Dec 66	F-105F	63-8354	354 TFS/355 TFW
355	14 Mar 67	F-105D	62-4325	469 TFS/388 TFW
356	05 Apr 67	F-105D	62-4395	44 TFS/388 TFW
357	06 Jul 67	F-105D	61-0136	354 TFS/355 TFW
358	03 Aug 67	F-105D	61-0139	333 TFS/355 TFW
359	03 Aug 67	F-105D	62-4240	333 TFS/355 TFW
360	07 Sep 67	F-105F	63-8260	13 TFS/388 TFW
361	26 Sep 67	F-105F	63-8267	357 TFS/355 TFW
362	03 Oct 67	F-105D	59-1824	354 TFS/355 TFW
363	05 Oct 67	F-105D	62-4329	354 TFS/355 TFW
364	06 Oct 67	F-105F	63-8272	44 TFS/388 TFW

Appendix D—F-105 Losses in Southeast Asia

Hit Location	Loss Location	Crew	Status
Engine Failure	Thailand		
Engine Failure	Thailand	Capt. R. Greskowiak	Died in Accident
Fuel Exhaustion	Thailand		
Engine Failure	Thailand		
Pilot Error	Thailand		
Engine Failure	Thailand		
Engine Failure	Thailand	Capt. W. N. Miller	Died in Accident
Engine Failure	Thailand		
Engine Failure	Thailand		
Engine Failure	Thailand		
Pilot Error	Thailand		
Engine Failure	Thailand		
Mid-Air Collision	Thailand		
Engine Failure	Thailand	Capt. J. A. McCurdy	Died in Accident
Ordnance System	Thailand		
Engine Failure	Thailand		
Flameout	Thailand		
Engine Failure	Thailand		
Engine Failure	Thailand	1Lt. G. F. Bulluck	Died in Accident
Crash on Takeoff	Thailand	Maj. D. W. Milliman	Died in Accident
Engine Failure	Thailand		
Engine Failure	Thailand		
Engine Failure	Thailand		
Control Failure	Thailand		
Drag Chute Malfunction	Thailand		
Engine Failure	Thailand		
Mid-Air Collision	Thailand	Capt. J. W. Bisschoff	Died in Accident
Mid-Air Collision	Thailand		
Engine Failure	Thailand	Col. McInerney/Capt. Shannon	Recovered
Hit tree during landing	Thailand		
Fuel Exhaustion	Thailand	Maj. R. R. King	
Engine Failure	Thailand		
Drag Chute Malfunction	Thailand		

F-105 Losses in Southeast Asia *(Continued)*

	Losses to Other than Combat (accidents, etc.)			
Loss #	Date	Aircraft	Serial No.	Unit
365	16 Oct 67	F-105D	60-0461	469 TFS/388 TFW
366	23 Oct 67	F-105D	61-0181	333 TFS/355 TFW
367	23 Oct 67	F-105D	62-4335	333 TFS/355 TFW
368	25 Oct 67	F-105D	59-1737	469 TFS/388 TFW
369	27 Oct 67	F-105D	61-0195	354 TFS/355 TFW
370	28 Oct 67	F-105D	62-4356	354 TFS/355 TFW
371	02 Jan 68	F-105D	61-0149	355 TFW
372	06 Jan 68	F-105F	62-4441	355 TFW
373	29 Jan 68	F-105D	60-0478	355 TFW
374	11 Mar 68	F-105D	60-0501	357 TFS/355 TFW
375	26 Mar 68	F-105D	60-0462	469 TFS/388 TFW
376	25 Apr 68	F-105D	60-0436	34 TFS/388 TFW
377	14 May 68	F-105D	61-0132	34 TFS/388 TFW
378	26 Jun 68	F-105D	58-1150	34 TFS/388 TFW
379	17 Aug 68	F-105D	61-0219	469 TFS/388 TFW
380	25 Aug 68	F-105F	63-8323	357 TFS/355 TFW
381	07 Sep 68	F-105F	63-8289	44 TFS/388 TFW
382	12 Sep 68	F-105D	62-4359	469 TFS/388 TFW
383	17 Nov 68	F-105D	61-0095	34 TFS/388 TFW
384	02 May 69	F-105F	62-4445	354 TFS/355 TFW
385	17 Jul 69	F-105D	62-4394	333 TFS/355 TFW
386	07 Oct 69	F-105D	62-4243	333 TFS/355 TFW
387	28 Nov 69	F-105D	60-0435	333 TFS/355 TFW
388	28 Nov 69	F-105D	61-0196	333 TFS/355 TFW
389	15 Apr 70	F-105G	62-4415	354 TFS/355 TFW
390	15 Apr 70	F-105D	61-0220	355 TFW
391	15 Nov 70	F-105G	63-8311	388 TFW
392	02 Feb 72	F-105G	63-8284	17 WWS/388 TFW
393	17 May 72	F-105G	63-8347	561 WWS/388 TFW

Hit Location	Loss Location	Crew	Status
Cocked Main Gear	Thailand		
Mid-Air Collision	Thailand		
Mid-Air Collision	Thailand	Capt. R. W. McClean	Died in Accident
Collided with C-123 while landing	South Vietnam	Maj. A. F. Britt	Died in Accident
Engine Failure	Thailand		
Engine Failure	Thailand		
Fuel Exhaustion	Thailand		
Engine Failure	Thailand		
Engine Failure	Thailand		
Engine Failure	Thailand		
Takeoff Abort—Right Main Tire Blew	Thailand		
Crashed on Landing	Thailand	Maj. B. R. Givens	Died in Accident
Mid-Air Collision	Thailand	Maj. S. R. Bass	Died in Accident
Structural Failure	Thailand		
Drag Chute Malfunction	Thailand	Capt. N. R. Koontz, Jr.	Died in Accident
Oil System Failure	Thailand		
Engine Failure	Thailand		
Fuel Exhaustion	Thailand		
Engine Failure	Thailand		
Oil System Failure	Thailand		
Blew Tire on Takeoff	Thailand	Maj. F. W. Shattuck, Jr.	Died in Accident
Fuel Exhaustion	Thailand		
Mid-Air Collision	Thailand		
Mid-Air Collision	Thailand		
Pilot Error	Thailand		
Material Failure	Thailand		
Engine Failure	Thailand		
Engine Failure	Thailand	Maj. C. H. Stone	Died in Accident
Blown Tire on landing	Thailand		

APPENDIX E

Thunderchief MiG Killers

			As listed in "Aces and Aerial Victories"					
No.	Date	Name	Pos.	Squadron/Wing	#	Type	Call Sign	Weapon
1	29 Jun 1966	Major Fred L. Tracy	Pilot	421TFS/388TFW	1	MiG-17	Crab 02	20-mm
2	18 Aug 1966	Major Kenneth T. Blank	Pilot	34TFS/388TFW	1	MiG-17	Honda 02	20-mm
3	21 Sep 1966	1st Lieutenant Karl W. Richter	Pilot	421TFS/388TFW	1	MiG-17	Ford 03	20-mm
4	21 Sep 1966	1st Lieutenant Fred A. Wilson	Pilot	333TFS/355TFW	1	MiG-17	Vegas 02	20-mm
5	04 Dec 1966	Major Roy S. Dickey	Pilot	469TFS/388TFW	1	MiG-17	Elgin 04	20-mm
7	10 Mar 1967	Captain Max C. Brestel	Pilot	354TFS/355TFW	2	MiG-17	Kangaroo 03	20-mm
8	26 Mar 1967	Colonel Robert R. Scott	Pilot	333TFS/355TFW	1	MiG-17	Leech 01	20-mm
9	19 Apr 1967	Major Leo K. Thorsness	Pilot	357TFS/355TFW	1	MiG-17	Kingfish 01	20-mm
—	19 Apr 1967	Captain Harold E. Johnson	EWO	357TFS/355TFW	1	MiG-17	Kingfish 01	20-mm
10	19 Apr 1967	Captain William E. Eskew	Pilot	354TFS/355TFW	1	MiG-17	Panda 01	20-mm
11	19 Apr 1967	Major Jack W. Hunt	Pilot	354TFS/355TFW	1	MiG-17	Nitro 01	20-mm
12	19 Apr 1967	Major Frederick G. Tolman	Pilot	354TFS/355TFW	1	MiG-17	Nitro 03	20-mm
13	28 Apr 1967	Lieutenant Colonel Arthur F. Dennis	Pilot	357TFS/355TFW	1	MiG-17	Atlanta 01	20-mm
14	28 Apr 1967	Major Harry E. Higgins	Pilot	357TFS/355TFW	1	MiG-17	Spitfire 01	20-mm
15	30 Apr 1967	Captain Thomas C. Lesan	Pilot	333TFS/355TFW	1	MiG-17	Rattler 01	20-mm
16	12 May 1967	Captain Jacques A. Suzanne	Pilot	333TFS/355TFW	1	MiG-17	Crossbow 01	20-mm
17	13 May 1967	Captain Charles W. Couch	Pilot	354TFS/355TFW	1	MiG-17	Chevrolet 03	20-mm
18	13 May 1967	Lieutenant Colonel Philip C. Gast	Pilot	354TFS/355TFW	1	MiG-17	Chevrolet 01	20-mm
19	13 May 1967	Major Robert G. Rilling	Pilot	333TFS/355TFW	1	MiG-17	Random 01	AIM-9
20	13 May 1967	Major Carl D. Osborne	Pilot	333TFS/355TFW	1	MiG-17	Random 03	AIM-9
21	13 May 1967	Major Maurice E. Seaver	Pilot	44TFS 388TFW	1	MiG-17	Kimona 02	20-mm

No.	Date	Name	Pos.	Squadron/Wing	#	Type	Call Sign	Weapon
22	03 Jun 1967	Major Ralph L. Kuster	Pilot	13TFS/388TFW	1	MiG-17	Hambone 02	20-mm
23	03 Jun 1967	Captain Larry D. Wiggins	Pilot	469TFS/388TFW	1	MiG-17	Hambone 03	AIM-9/20-mm
24	23 Aug 1967	1st Lieutenant David B. Waldrop	Pilot	34TFS/388TFW	1	MiG-17	Crossbow 03	20-mm
25	18 Oct 1967	Major Donald M. Russell	Pilot	333TFS/355TFW	1	MiG-17	Wildcat 04	20-mm
26	27 Oct 1967	Captain Gene I. Basel	Pilot	354TFS/355TFW	1	MiG-17	Bison 02	20-mm
26.5	19 Dec 1967	Major William M. Dalton	Pilot	333TFS/355TFW	0.5	MiG-17	Otter 02	20-mm
—	19 Dec 1967	Major James L. Graham	EWO	333TFS/355TFW	0.5	MiG-17	Otter 02	20-mm
27.5	19 Dec 1967	Captain Phillip M. Drew	Pilot	357TFS/355TFW	1	MiG-17	Otter 03	20-mm
—	19 Dec 1967	Major William H. Wheeler	EWO	357TFS/355TFW	1	MiG-17	Otter 03	20-mm

INDEX

References to photos or illustrations are followed by a (Illus.) indication.

2.75-inch FFAR, 5, 54 (Illus.), 123
2nd AD, 111
4th ATAF, 50
4th TFW, 25, 27, 29, 31, 33, 65
7th AF, 95
8th TFW, 95
12th TFS, 86, 135
13th Raiders (*see* 13th TFS)
13thTFS, 99, 122
17th WWS, 140 (Illus.)
23rd TFW, 33, 34, 58, 59, 73
34th TFS, 80
35th TFW, 141 (Illus.), 146 (Illus.)
36th TFW, 50, 73
41st AD, 97–98, 100
44th TFS, 73, 76 (illus.), 83 (Illus.), 102, 128 (Illus.)
49th TFW, 50
57th FWW, 125, 144 (Illus.)
66th FWS, 125, 147 (Illus.), 148
67th TFS, 73
108th TFW, 33, 42
113 TFG, 107
141st TFS, 33
184th TFTG, 107
192nd TFG, 107
2098-aircraft (*see* Ryan's Raiders)
301st TFW, 107
333rd TFS, 76 (Illus.), 171 (Illus.)
334th TFS, 78, 86
335th TFS, 25, 52
355th TFW, 75 (Illus.), 76 (illus.), 78, 80, 82, 84 (Illus.), 86, 94, 96, 103–104, 107, 122–123
357th TFS, 77 (Illus.)
388th TFW, 76 (Illus.), 78, 82, 95–97, 103, 107, 122–124, 129
441st A&E, 97
457th TFS, 58, 59
466th TFS, 33 (Illus.), 34 (Illus.), 35 (Illus.)
469th TFS, 86, 122 (Illus.)

507th TFG, 107
508th TFG, 33
561st TFS, 34 (Illus.), 136, 141 (Illus.)
562nd TFS, 73, 146 (Illus.)
563rd TFS, 73, 74 (Illus.), 107 (Illus.)
4500th CCTW, 33
4520 CCTW, 50, 65, 66 (Illus.)
6010th WWS, 86, 135, 139 (Illus.), 140 (Illus.)
6234th TFW(P), 74
6235th TFW(P), 74

AE-100 (ATI) (*see* Az-El)
AIM-9 Sidewinder, 5, 23, 172 (Illus.), 173
  dual launcher, 175 (Illus.)
Air America, 59
Air Force Museum (*see* United States Air Force)
AGM-12 Bullpup, 53, 73–74, 77, 84, 86, 173 (Illus.), 174 (Illus.)
AGM-45 Shrike, 86, 115, 118, 122 (Illus.), 126 (Illus.), 127 (Illus.)
  F-105 designated to carry, 56, 77
  ADU-315/316 dual adapters, 135, 136 (Illus.), 138 (Illus.), 139 (Illus.)
  deficiencies, 122, 128–129
  emergency Shrike effort, 128–129
  LAU-34/A launcher, 135
AGM-78 Standard ARM, 126 (Illus.), 128 (Illus.)
AGM-78A (Mod 0), 129–130
AGM-78B/C (Mod 1), 129, 133
  Improvements over Shrike, 129–130
Aircraft and Weapon Board (*see* United States Air Force)
ALE-2, 11, 23, 46, 48, 53, 175 (Illus.)
ALQ-31, 23, 48, 61, 93
ALQ-51 (Sanders Associates), 114–115, 125
ALQ-55 (Sanders Associates), 103
ALQ-59 (Hallicrafters), 103, 104 (Illus.)

ALQ-71 (GE/Hughes), 94 (Illus.), 95–96, 114, 125
  initial problems in Southeast Asia, 95
ALQ-87 (GE/Hughes), 96, 104 (Illus.), 125
ALQ-101 (Westinghouse), 96, 125–126
ALQ-105 (Westinghouse), 124 (Illus.), 126, 135 (Illus.)
ALQ-119 (Westinghouse), 96
ALR-31 (Loral), 131 (Illus.), 149
ALR-46 (Dalmo-Victor/Itek), 131 (Illus.), 134 (Illus.), 148–149
ALT-34 (Borders), 126, 133
Anderson, Col. Clarence E. "Bud," 84
antenna locations, 60 (Illus.), 130 (Illus.)
Anti-SAM seminar, 92
Anti-SAM Tactics and Capability Program, 113–114
Anti-SAM Task Force, 112–113
AP-44A (*see* Republic Aircraft Corporation)
AP-63 (*see* Republic Aircraft Corporation)
AP-63-FBX (*see* Republic Aircraft Corporation)
APN-105 Doppler, 19–21, 39, 50, 166
APN-131 Doppler, 59, 167
APR-23B (Melpar), 114
APR-25 (ATI), 76 (Illus.), 78 (Illus.), 85 (Illus.), 93, 114–116, 119–120, 123, 126
APR-26 (ATI), 93, 114–115, 118, 120, 123, 126
APR-35 (ATI), 126, 129 (Illus.), 132 (Illus.), 133, 137
APR-36 (ATI), 126, 133, 148
APR-37 (ATI), 121, 126, 133, 148
APR-38, 126, 149
APS-54, 11, 32 (Illus.), 39, 48, 53
APS-92, 46
APS-107 (XA-1) (Bendix), 113–114
APS-107B (X-1) (Bendix), 114–115
ARC-51 (Admiral), 89–90

223

area-rule principle, 7, 8, 157
ARN-92 LORAN, 58, 169
arresting hook, 163 (Illus.)
ASG-19, 46, 49–50, 53, 58, 78, 97, 167, 183–185
   use of T-39 trainers, 50, 64
   all-weather bombing system, 97–102
Az-El
   AE-100 (ATI), 120
   Pointer III (AEL), 120

BASS I/II (LTV), 127–128
battle damage, 88, 90 (Illus.), 91 (Illus.), 92 (Illus.), 112
Beaird, Henry G. Jr., 19
Bell, Maj. Kenneth, 145
Bennett, Maj. W. G., 92, 210
blind bombing mod, 97–102
bomb damage assessment camera, 84, 85 (Illus.), 129 (Illus.)
bomb damage assessment receiver, 133–134
Boyd, General Albert, 4
Brestel, Captain Max C., 80, 223
Brookley AFB, 33, 50
   (see also Mobile Air Materiel Command Area)
Broughton, Jack, 92
Bubbles I, 85 (Illus.)
buddy bombing, 89 (Illus.)
buddy refueling, 10 (Illus.)
Bullpup (see AGM-12 Bullpup)
buzz numbers, 5

Cagle, V.Adm. Malcolm W., 112
cameras, 35
camouflage paint, 86–87, 89 (Illus.), 168 (Illus.), 169
chaff dispensers (see ALE-2)
Chairsell, Gen. William S., 95
climatic evaluations, 21, 22 (Illus.)
"Colonel Computer" (see ALQ-59 (Hallicrafters))
Combat Martin, 103 (Illus.), 104
Combat Skyspot, 75, 102, 104–105
Combat Target, 96
Commando Club, 105, 106 (Table)
Commando Nail, 100
Commando Nail Papa, 100, 102
Cook, General Orval R., 5
Cook-Craigie concept of "concurrency," 5, 21
   application of, 5
   failure of, 27–28
Craigie, General Laurence C., 5
Crossfield, A. Scott, 14
Cuban missile crisis, 33
Cutter, Capt. J. D. , 140, 216

Davis-Monthan AFB , 59, 107, 149, 150 (Illus.)

Dempster, Brig. Gen. Kenneth C., 112
Dethlefsen, Capt. Merlyn Hans, 143–145, 148
Devlin, Captain Eugene, 42
"Dixie Twister," 78 (Illus.)
Doppler navigation (see APN-105)
DPN-61 (Bendix), 49–50, 114–115, 117 (Illus.)

EC-121 Rivet Top, 127
ECM pods (see ALQ-71; ALQ-87; ALQ-101; ALQ-105)
Edwards AFB, 6, 19, 21, 25, 30
EF-105F, 103, 125
Eglin AFB, 20, 25, 67 (Illus.), 93, 107, 118, 136
   climatic laboratory, 21 (Illus.)
ejection seat, 27–28, 62 (Illus.), 65 (Illus.), 157, 158 (Illus.)
   supersonic, 56 (Illus.)
Electronic Intelligence, 127
ELINT (see Electronic Intelligence)
ER-142 (ATI), 97 (Illus.), 98, 123, 125, 133, 135
ER-151 (ATI), 127
ER-168 (ATI) (see APR-35 (ATI))

F-84 (Republic), 1, 2
F-84X (Republic), 2
F-100B (North American) (see YF-107A (North American))
F-105A, 2–4
   cutting program back, 4
F-105B, 5, 6, 9, 17–35, 46
   cost of, 29
   first flight, 19
   first production aircraft, 27
   first public display, 20
   end of production, 30
   evaluation report, 21–22
   maintenance man-hours, 29
   shutting down production line, 27
   Thunderbirds, 39–42
   total cost per aircraft, 30
F-105B(AW) (see F-105D)
F-105C, 17, 19, 38–39, 61
   cockpit mockup, 38 (Illus.), 63
F-105D, 20, 45–64
   cockpit mockup, 47
   cost of, 58
   differences from F-105B, 45–46, 177 (Illus.)
   first flight, 49
   groundings, 56
   labor strike halts production, 53
   losses in Southeast Asia, 86, 106–107
   maintenance man-hours, 58
   modifications in Southeast Asia, 86, 87 (Illus.), 88, 160, 165 (Illus.), 178
   proposed FAI record flight, 69
   T-Stick II (see Thunderstick II)

F-105E, 20, 45, 47–48, 61 (Illus.), 62–64
   cockpit mockup, 62 (Illus.), 63
   cost delta from F-105D, 47
   use of material for F-105D, 63
F-105F, 64–68
   back seat capabilities, 155–156
   blind bombing mod, 97–102
   cockpit preview, 65
   cost of, 67–68
   ejection system test, 65 (Illus.)
   last production aircraft, 67
   maintenance man-hours, 68
   modifications in Southeast Asia, 86, 87 (Illus.), 88, 160, 165 (Illus.), 178
F-105F Wild Weasel (see Wild Weasel)
F-105G (proposed), 68
F-105G Wild Weasel (see Wild Weasel)
F-105H (proposed), 68
Fan Song radar (see SA-2 surface-to-air missile)
fatigue tests, 30, 53
FFAR (see 2.75-inch FFAR)
Fire Can radar, 92–93, 96, 118, 123–124
flutter concerns, 9
folding fin aircraft rocket (see 2.75-inch FFAR)
Foley, Capt. Peter, 86
Fraser, Capt. K. J., 140, 216
Frederick, Capt. W. V. , 90, 210
French interest in F-105, 69

GAM-83 Bullpup (see AGM-12 Bullpup)
General Operating Requirement (GOR)
   GOR-49, 5
   GOR-49-1, 21, 45, 183
   GOR-49-2, 59–60
   GOR-166, 21
Gilroy, Capt. Kevin "Mike," 143–145
Graben, Lt. Dave, 73
"Great Pumpkin, The," 83
Gulf of Tonkin incident, 73

Haiphong, 75
"Half A Yard," 128
Hanoi, 75, 77
Hendrix, Lindell, 36, 49
HRB-Singer 934-1B, 121
Hoza, Lt. Col. Paul, 52
Huey, Sidney R., 6, 7 (Illus.)
hunter-killer, 113, 122, 137

initial operational capability
   F-105A, 2, 3, 4, 5
   F-105B, 23
   F-105D, 50
IR-133 (ATI), 113, 116, 118–120, 123
Iron Hand, 112, 115, 123
   attacks, 95, 112, 122, 136
   SAM site location techniques, 111–112, 120

J57 (Pratt & Whitney), 4, 8, 11, 173
J67 (Wright), 4
J71 (General Motors Allison), 2, 4
J73 (General Electric), 2
J75 (Pratt & Whitney), 4, 5, 12, 17, 19, 28–30, 36, 45, 49, 53, 54, 55 (Illus.), 61, 80 (Illus.), 161, 169 (Illus.), 174, 178 (Illus.)
  JT4B variant, 68–69
JF-105B, 21, 28, 35–37
  first flight, 36
  official designation, 43 (note 50)
  modify to F-105E configuration, 61
Johnson, Capt. Harold E., 145–148

Kartiveli, Alexander, 2
Korat RTAFB, 57, 73–74, 86, 95, 120–122, 136
KS-24A/KS-27, 4, 172–173
Kuster, Maj. Ralph L. Jr., 100, 109 (note 121)

Langley Aeronautical Laboratory (*see* National Advisory Committee on Aeronautics)
Last Flight, 151 (Illus.)
Linebacker (*see* Operation Linebacker)
Loftin, Larry, 9
Luftwaffe interest in F-105, 68

M39A1 20-mm cannon, 35
M61 20-mm cannon, 4, 5, 26 (Illus.), 46 (Illus.), 79 (illus.), 170 (Illus.), 177 (Illus.)
MA-8, 12, 20, 25, 29–30, 38, 46, 78, 166 (Illus.), 181–185
MA-12, 12
McClellan AFB, 114, 118–119
McClelland, Maj. William, 90
McConnell, Gen. John P., 114
McCoy AFB, 33
McNamara, Robert, 64, 111
Medal of Honor, 143–148
"Memphis Belle II," 77 (Illus.)
MiG Fighters
  comparison of, 78–80
  F-105s shot down by, 74
  shooting down of, 80, 223–224
"Miss Universe," 88 (Illus.)
MOAMA (*see* Mobile Air Materiel Command Area)
Mobile Air Materiel Command Area (MOAMA), 33, 57
Momyer, Gen. William, 93
Moore, Capr. H. W. Jr., 82, 212
Moore, Brig. Gen. Joseph H., 30, 100
MSQ-77, 104–105

NASARR (*see* R-14 North American)
National Advisory Committee on Aeronautics, 7–10
  High-Speed Flight Station, 13–14
  Langley Aeronautical Laboratory, 7–9

nuclear weapons, 3, 12, 17, 23, 24 (Illus.), 52, 53

"Old Crow II," 84 (Illus.)
Olds, Col. Robin, 95
Operation Bullseye, 96
Operation Coronet Bolo I, 141
Operation Coronet Exxon, 141
Operation Flaming Dart, 74
Operation Linebacker I, 86
Operation Linebacker II, 86, 103, 137, 142
Operation Northscope, 97 (illus.), 97–102
Operation Rolling Thunder, 74–75, 88, 111, 132, 136, 137

Pointer III (AEL) (*see* Az-El)
Project Back Bone, 42
Project Big Bear, 31
Project Black Box, 49–50
Project Fast Wind, 30
Project Flying Fish, 52, 157
Project GUNVAL, 4
Project High Flight, 50
Project Look Alike, 31, 33, 52 (Illus.), 54, 66 (Illus.), 89 (Illus.), 169
Project Optimize, 29–30
Project Skyspot, 95
Project Stay On Top III, 53
Project Vampyrus, 94
Proud Deep Alpha, 136
Prong Tong, 112

QRC-128 (Hallicrafters) (*see* ALQ-59 (Hallicrafters))
QRC-160, 93–94
QRC-160A (GE/Hughes) (*see* ALQ-71 (GE/Hughes))
QRC-160-1 (GE/Hughes) (*see* ALQ-71 (GE/Hughes))
QRC-160-8 (GE/Hughes) (*see* ALQ-87 (GE/Hughes))
QRC-207 Fan Song simulator, 121
QRC-288, 126
QRC-301 (Sanders Associates), 125
QRC-317 (NAA), 121, 125
QRC-321 (Westinghouse), 124 (Illus.), 125, 135 (Illus.)
QRC-333, 127
QRC-335 (Westinghouse) (*see* ALQ-101 (Westinghouse))
QRC-339 (ATI) (*see* ER-151 (ATI))
QRC-373 (Borders) (*see* ALT-34 (Borders))
QRC-380 (Westinghouse) (*see* ALQ-105)
QRC-535 (Itek), 148
QRC-555 (Itek), 148

R-14 (North American), 12, 20, 46, 59, 61, 63–64, 167, 183
  all-weather bombing system, 97–102
  R-14K T-Stick II, 59, 169

Radar Homing and Warning (RHAW), 90
  APR-25 (*see* APR-25 (ATI))
  APR-26 (*see* APR-26 (ATI))
  APR-36 (*see* APR-36 (ATI))
  APR-37 (*see* APR-37 (ATI))
  Vector IV (*see* Vector IV)
  WR-300 (*see* WR-300 (ATI))
radio relay system (*see* ARC-51 (Admiral))
ram air turbine, 164, 171 (Illus.)
Red River Valley, 77, 132
Republic Aviation Corporation, 1, 116, 118–119
  AP-44A (*see* XF-103)
  AP-63, 3
  AP-63-5, 38
  AP-63-33, 61
  AP-63-36, 69
  AP-63-FBX, 2, 3 (Illus.)
  AP-95 missile, 24
  ES-347-1 (F-105C), 38
  ES-349 (F-105B), 17
  ES-350 (RF-105B), 35
  proposal to reopen production line, 70
  RF-105, 4, 19, 35
  RF-105B, 5, 17, 28, 35–37
  RF-105D, 59–61
RHAW (*see* Radar Homing and Warning (RHAW))
Richter, 1Lt. Karl, 86, 223
Rolling Thunder (*see* Operation Rolling Thunder)
Roth, Russell "Rusty" M., 6
Route Packs, 75
  Route Pack 1, 80, 100-102, 105, 121
  Route Pack 5, 99, 101
  Route Pack 6, 75, 121
  Route Pack 6A, 75, 99, 101
Ryan, Gen. John D., 96–97
Ryan's Raiders, 96, 97 (Illus.) 97–102; (*see also* Commando Nail; Operation Northscope)
Royal Air Force interest in F-105, 69
RTAFB (*see* Korat RTAFB); (*see also* Takhli RTAFB)

SA-2 surface-to-air missile, 90, 95, 113
  description, 90, 92
  F-105 kills, 86
  Fan Song radar, 90, 92–93, 95–96, 111, 118, 120, 123, 129, 137
  Fan Song simulators, 121
  mobility, 112
  site construction, 90, 111
  T-8209/Team Work radar, 137
Sacramento Air Materiel Area (SMAMA), 57, 59, 114–115, 118, 120–121, 123, 133–134, 144 (Illus.)
SAM (*see* SA-2 surface-to-air missile)
SEE SAMS (APGC/NAA), 120–121, 125–126
SEE SAMS B, 121

Seymour-Johnson AFB, 25, 31, 78
Scott, Colonel Robert R., 25
Sidewinder (see AIM-9)
Site 85 (TSQ-81), 105
"Sittin Pretty," 82 (Illus.)
SMAMA (see Sacramento Air Materiel Area (SMAMA))
Son Nhut, 73
Spearman, Leroy, 8
special stores (see nuclear weapons)
speed brakes, 3, 17, 18 (Illus.), 178 (Illus.)
Spoon Rest radar, 126
Spot SAM, 126
SST-181X, 47 (Illus.), 104
strike camera (see bomb damage assessment camera)

T-130 0.60-caliber gun, 2, 4
T-160 20-mm cannon (see M39A1 20-mm cannon)
T-171 20-mm cannon (see M61 20-mm cannon)
T-39 trainers (ASG-19), 50
T-Stick II (see Thunderstick II)
TAC (see United States Air Force)
Tactical Air Command (see United States Air Force)
Tactical Air Warfare Center (see United States Air Force)
Takhli RTAFB, 73–75, 121–122
TAWC (see United States Air Force)
TF-105B, 43 (note 50)
Thaete, Capt. D. K., 82, 214
"The Mercenary," 82 (Illus.)
Thorsness, Maj. Leo K., 145–148
Thud
  definition of name, 1
  origin of name, 56
Thud Ridge, 77
Thunderbirds, 39–42
  accident, 42
Thunderchief
  approval of name, 19
Thunderstick (see ASG-19)
Thunderstick II, 57 (Illus.), 58–59, 60 (Illus.), 169
TSQ-81, 105–106
Treyz, Lt. Col. Fred A., 99

United States Air Force
  Air Training Command, 17, 38
  Aircraft and Weapon Board, 2
  Museum, 14
  Tactical Air Command (TAC), 35
  Tactical Air Warfare Center (TAWC), 119
  Wright Air Development Center (WADC), 4, 25, 35

Vector IV (ATI), 113–114, 116
vertical tape instruments, 46, 47 (Illus.), 49

WADC (see United States Air Force)
Waldrop, Lt. David B., 81 (Illus.), 223
Waller, Maj. Larry, 86
Walsh, Maj. Francis P., 101
Whitcomb, Richard, 6
Wild Weasel, 77, 80, 81, 107, 111–149
  Aircraft
    F-100F, 93, 114, 116, 118
    F-105D, 50, 113–115
    F-105F, 116, 118–124, 137
    F-105F modification problems, 118–119
    EF-105F, 103, 125
    F-105G, 125, 133 (Illus.), 134, 148
    Ryan's Raiders, 97–102
  Avionics
    ALQ-51 (Sanders Associates) (see ALQ-51 (Sanders Associates))
    ALQ-71 (GE/Hughes) (see ALQ-71 (GE/Hughes))
    ALQ-87 (GE/Hughes) (see ALQ-87 (GE/Hughes))
    ALQ-101 (Westinghouse) (see ALQ-101 (Westinghouse))
    ALQ-105 (Westinghouse) (see ALQ-105 (Westinghouse))
    ALR-31 (Loral) (see ALR-31 (Loral))
    ALR-46 (Dalmo-Victor/Itek) (see ALR-46 (Dalmo-Victor/Itek))
    ALT-34 (Borders) (see ALT-34 (Borders))
    APR-23B (Melpar) (see APR-23B (Melpar))
    APR-25 (ATI) (see APR-25 (ATI))
    APR-26 (ATI) (see APR-26 (ATI))
    APR-35 (ATI) (see APR-35 (Itek))
    APR-36 (ATI) (see APR-36 (ATI))
    APR-37 (ATI) (see APR-37 (ATI))
    APR-38 (see APR-38)
    APS-107 (XA-1) (Bendix) (see APS-107 (XA-1) (Bendix))
    APS-107B (X-1) (Bendix) (see APS-107B (X-1) (Bendix))
    Az-El/AE-100 (ATI) (see Az-El)
    Az-El/Pointer III (AEL) (see Az-El)
    BASS I/II (LTV) (see BASS I/II)
    bomb damage assessment receiver (see bomb damage assessment receiver)
    DPN-61 (Bendix) (see DPN-61 (Bendix))
    ER-142 (ATI) (see ER-142 (ATI))
    ER-151 (ATI) (see ER-151 (ATI))
    ER-168 (ATI) (see APR-35 (Itek))
    HRB-Singer 934-1B (see HRB-Singer 934-1B)
    IR-133 (ATI) (see IR-133 (AIT))
    Maxson fin cap RHAW system (see Mason fin cap RHAW system)
    QRC-160-1 (GE/Hughes) (see ALQ-71 (GE/Hughes))
    QRC-160-8 (GE/Hughes) (see ALQ-87 (GE/Hughes))
    QRC-288 (see QRC-288)
    QRC-301 (Sanders Associates) (see QRC-301 (Sanders Associates))
    QRC-317 (NAA) (see QRC-317 (NAA))
    QRC-321 (see QRC-321)
    QRC-333 (see QRC-333)
    QRC-335 (Westinghouse) (see ALQ-101 (Westinghouse))
    QRC-339 (ATI) (see ER-151 (ATI))
    QRC-373 (Borders) (see ALT-34 (Borders))
    QRC-535 (Itek) (see QRC-535 (Itek))
    QRC-555 (Itek) (see QRC-555 (Itek))
    SEE SAMS (APGC/NAA) (see SEE SAMS (APG/NAA))
    SEE SAMS B (see SEE SAMS B)
    Vector IV (ATI) (see Vector IV)
    WR-300 (ATI) (see WR-300 (ATI))
  losses in Southeast Asia, 86
  projects
    Wild Weasel I, 114–116, 118, 120
    Wild Weasel IA, 50, 115–116, 117 (Illus.)
    Wild Weasel II, 113–115
    Wild Weasel III, 116, 118–137
WR-300 (ATI), 113–114
Wright Air Development Center (WADC) (see United States Air Force)
WS-204A (see XF-103)
WS-306 (see F-105A)
WS-306L (see F-105C)

XF-103, 1, 8

YF-105A, 5, 6, 7 (Illus.), 8 (Illus.), 21, 159, 173
  differences between F-105B, 18 (Illus.), 162, 174
  first flight, 5
  transport by C-124 to Edwards, 5
YF-107A (North American), 11–14, 46
  first flight, 12